Adaptive Software Development

A COLLABORATIVE APPROACH TO MANAGING COMPLEX SYSTEMS

Adaptive Software Development

A COLLABORATIVE APPROACH TO MANAGING COMPLEX SYSTEMS

James A. Highsmith III

foreword by Ken Orr

Dorset House Publishing
353 West 12th Street
New York, NY 10014

Library of Congress Cataloging-in-Publication Data

Highsmith, James A.
    Adaptive software development : a collaborative approach to managing
complex systems / James A. Highsmith, III ; foreword by Ken Orr.
        p. cm.
    Includes bibliographical references and index.
    ISBN 0-932633-40-4 (softcover)
        1. Computer software--Development. 2. Computer Systems--Management.
    3. Management information systems. I. Title.

QA76.76.D47 H55 1999
658'.0551--dc21   99-052212

Graphics from TASK FORCE Clip Art appear in Figs. 1.1, 1.2, and 9.6 with permission from New Vision Technologies. Copyright © 1998. All rights reserved.

Cover Design: David McClintock
Cover Photograph: Stuart Ruckman Photography
Author Photograph: Robert Munk Photography

Copyright © 2000 by James A. Highsmith III. Published by Dorset House Publishing Co., Inc., 353 West 12th Street, New York, NY 10014.

Distributed in the English language in Singapore, the Philippines, and Southeast Asia by Alkem Company (S) Pte. Ltd., Singapore; in the English language in India, Bangladesh, Sri Lanka, Nepal, and Mauritius by Prism Books Pvt., Ltd., Bangalore, India; and in the English language in Japan by Toppan Co., Ltd., Tokyo, Japan.

Printed in the United States of America

Library of Congress Catalog Number: 99-052212

ISBN: 0-932633-40-4   12   11   10   9   8   7   6   5   4

Dedication

To William Byron Mowery, whose writing influenced me in subtle ways I am just beginning to understand. Bill Mowery wrote several dozen books and published hundreds of articles from the 1920's through the early 1950's. He taught creative writing at New York University. Although I never met him, he was my grandfather, and he passed on a love and understanding of writing to my mother, Dodie, who is an accomplished writer herself. I am indebted to her for a lifetime of encouragement, and to him for a legacy I've always wanted to contribute to in my own way.

Contents

Acknowledgments

The book before you is a result of creative collaboration. While I take full responsibility for the content, its quality would be greatly diminished without the efforts of my colleagues and clients—the people who have contributed in a variety of ways toward the book's completion.

Ken Orr, with whom I have worked for nearly twenty years. I have always been in awe of Ken's ability to conceptualize and create visions of the possible. Over the years, my relationship with Ken has spanned that of customer, employee, coworker, and colleague—but most importantly, friend.

Lynne Nix, with whom I also have worked for many years. Lynne is not only one of the best project managers I have ever known, but she is *the* best at conveying her knowledge to others. A number of the project management ideas in this book originated with Lynne.

Jerry Weinberg, whose writing has influenced me since the early 1970's. I've known Jerry since the mid-1980's and have always been inspired by his emphasis on and insight into the human side of our technological world.

Sam Bayer, who co-developed many of the RAD practices in this book and with whom I've worked on a number of accelerated development projects. As in any good collaborative effort, distinctions as to which of the ideas were mine and which were Sam's have long since been lost. Sam has an uncanny ability to extract the essence of a complex situation and communicate that essence in a cogent, simple way. His comments on draft versions of this book helped me focus more clearly.

Jerry Gordon, whose ideas impacted this book in many ways. I first met Jerry nearly twenty years ago when I was doing "structured stuff." Jerry introduced me to mountaineering and climbing. Many of the climbing stories in the book are based on trips he and I took. During the years we have worked together (on some form of business or another), Jerry was actually willing to *try* some of my ideas about adaptive development when they were still nascent.

Steve Smith, who read this manuscript more times, and in more different incarnations of it, than anyone else. His comments and incisive questions always expanded my thinking and were greatly appreciated. Steve and I share two joys—discussing difficult issues and taking long, arduous hikes.

Others who have contributed directly, and sometimes indirectly, include Cheryl Allen, Jim Davis, Rob Arnold, James Bach, Karen Coburn, Tom DeMarco, Anne Farbman, Dan Larlee, Adele Goldberg, Warren Keuffel, Martyn Jones, Steve McMenamin, Lou Russell, Robert Charette, Larry Proctor, George Engleberg, James Odell, Bruce Watson, George Johnson, and Wayne Collier. I thank them all for the wisdom they have so generously shared.

I had support from many friends in the Salt Lake climbing community who were party to the analogies used in the book. Among the many, I especially thank Amy Irvine, who first taught me that movement skills were more important than strength, and Doug Hunter, who helped me refine those skills. Thanks also to Mac Lund, my great friend and regular climbing, skiing, and hiking partner, and to Bob Richards, Brian Mecham, Dale Goddard, and the crew at Rockreation.

To my wife, Wendie, who encouraged me throughout the ups and downs of the writing process and who put up with my sleepless nights and pirated weekend mornings, I am profoundly grateful. To daughters Nikki and Debbie, and to the *other* Jim, my father, I am thankful for their support and encouragement.

I never truly understood the relationship between authors and publishers—now I do. The staff at Dorset House that worked with me on this book is exceptional. Mike Lumelsky, Matt McDonald, Bob Hay, David McClintock, and Wendy Eakin were all part of the collaborative effort that turned a manuscript into a polished product. My thanks to all of them.

Permissions
Acknowledgments

p. xxx: Material from Warren Keuffel, "People-Based Processes: a RADical Concept." *Software Development* (November 1995), p. 37. Reprinted with permission.

pp. 3-26: Material adapted from Jim Highsmith, "Software Ascents." *American Programmer* (June 1992), pp. 20-26. Reprinted with permission.

pp. 8; 37; 285; 286: Material from Stuart Kauffman, *At Home in the Universe*. Copyright © 1995 by Stuart Kauffman, pp. vii; 15; 100; 84. Reprinted by permission of Oxford University Press, Inc., and Brockman, Inc. All rights reserved.

p. 9: Reprinted from W. Brian Arthur, "Increasing Returns and the New World of Business." *Harvard Business Review* (July-August 1996), pp. 100, 101, 102, 107. Reprinted with permission.

pp. 9ff.: Material adapted from Jim Highsmith, "Messy, Exciting, and Anxiety-Ridden: Adaptive Software Development." *American Programmer* (April 1997), pp. 23-29. Reprinted with permission.

pp. 10; 329: Material from George Johnson, *Fire in the Mind: Science, Faith, and the Search for Order*. Copyright © 1995, pp. 235; 130. Reprinted by permission of Vintage Books, Random House, Inc. All rights reserved.

pp. 11; 58; 59; 143; 202; 205: Reprinted from Arie de Geus, *The Living Company: Habits for Survival in a Turbulent Business Economy*. Boston: Harvard Business School Press, 1997, pp. 11; 155; 46; 157; 140; 3. Reprinted with permission.

Foreword

I f I could, I would give a copy of Jim Highsmith's book to everyone involved in developing large systems—end users, managers, IT professionals, and most especially IT project managers! Jim's message is simple but vitally important: Large information systems don't have to take so long, they don't have to cost so much, and they don't have to fail. Unfortunately, as simple as Jim's message is, making it happen is an enormously difficult undertaking in most large organizations.

While many project management books deal extensively with the need for building systems faster in today's business workplace, Jim's is the first book I've read that addresses what must happen when management is faced both with the need for high-speed delivery and with business requirements that are rapidly changing as well!

Jim's solution to both these problems is straightforward—a radical form of incremental development. But actually developing large systems incrementally is a considerably more difficult business than just talking about it. Historically, large organizations tend to attack all problems by breaking them into pieces and assigning each of the pieces to different organizational units for parallel development. Management then places its faith in managing to predefined budgets and schedules, but usually does so without any clear idea of what exactly the completed project will produce or how the major pieces, the sub-projects, ultimately will fit together.

Jim's approach combines the best features of techniques that have been used piecemeal for a long time: customer focus groups, versioning, time-boxed management, and active prototyping. Used individually, these approaches can be effective; combined, they are dynamite.

In my own consulting work, I am often called in to turn off the life support on large, failed projects. Most of these systems fail because they lack the right systems strategies and because they take too long. Projects lasting three-to-five years are rarer and rarer these days because organizations know that they have to implement systems earlier to meet changing business needs. On the other hand, unstructured, short-term projects are a maintenance nightmare. So, since large-scale projects won't simply go away, they need to be approached in a different fashion. In order for organizations to develop large systems more successfully, they must develop incremental implementation plans and incremental architectures, and they must begin to build individual subprojects in small, short-term pieces. Jim Highsmith's book provides the framework for doing just that.

August 1999   Ken Orr
Topeka, Kansas

Preface

George Johnson's *In the Palaces of Memory* weaves a fascinating story about one of the most complex of all biological phenomena: human memory. From Johnson's stories of neurobiologists who study the neuron and its components—axons, synapses, dendrites—and try to model how neural cells excite each other to create patterns, to tales of computer scientists whose explorations of the mind use neural networks and artificial intelligence, the reader is drawn to the conclusion that how memory works is still shrouded in mystery.

Software development may be as close to a purely *mental* activity as any complex business undertaking. Just as we are puzzled by the workings of the mind, we are caught up in the enigma of how software products emerge from the minds of their creators.

Software seems so simple. A few operators, a few operands, and *voilà*—a program is created. Combinatorial mathematics, however, insures that programs of any length have infinite potential variety—perpetual novelty.

Chess also seems simple. With fewer than two dozen rules and despite several hundred years of play, chess's capacity to generate perpetual novelty has not diminished.

Perpetual novelty is one measure of complexity. However, *complex* is not the same as *chaotic*. Chaos is random. Complexity contains patterns—patterns that peek through the perpetual novelty and can be used by people in their struggle to thrive in our world. This book is about the complex human endeavor of building computer software. It is about using the emerging field of complexity science (or, more specifically, the field of complex adaptive systems theory) to aid us in our pursuit of ever-more-complex software products. At its core, software development may be about how we turn *memories* of our world into computer models.

Neurons are pieces that create the fabric of memory, but how this fabric is woven is unknown. However, not understanding the weaving does not keep us from using the fabric. The infinite fascination of chess arises from simple rules. We cannot *predict* where a chess game will go, but we can learn *patterns* of play that bring success. Both chess and memory provide instances of a phenomenon that in the language of complex adaptive systems is called "emergence." Emergence is a property of complex adaptive systems in which the interaction of the parts creates some greater property of the whole that cannot be fully explained from measured behaviors of the agents. These results come from exploiting patterns. Harnessing emergence is a major theme of this book.

Goals of the Book

Adaptive Software Development has five primary goals. Detailed in the paragraphs below, the goals define the essence of building better software in a world where high speed, change, and uncertainty are key characteristics of its intensifying complexity.

The first goal is to offer an alternative to the belief that optimization is the only solution to increasingly complex problems. Optimizing cultures believe they are in control, that they can impose order on the uncertainty around them. Imposed order is the product of rigorous engineering discipline and deterministic, cause-and-effect-driven processes. The alternative idea is one of an *adaptive* culture or mindset, of viewing organizations as complex adaptive systems, and of creating emergent order out of a web of interconnected individuals. An adaptive approach raises the acknowledgment of an uncertain and complex world from being a manager's death knell to being part of a recognized and accepted strategy.

The second goal is to offer a series of *frameworks* or *models* to help an organization employ adaptive principles. The Adaptive Development Life Cycle, for example, provides a framework that reinforces the concepts and details a practical way of moving from the conceptual to the actionable. While there are other types of iterative life cycles, they have not been based on underlying adaptive concepts but rather on short-cycle determinism. There is a certain synergy in linking iteration with adaptive ideas to combat complexity. The frameworks are supplemented by a discussion of various techniques (customer focus groups, for example). However, the focus is on the frameworks, not on a compendium of techniques.

The third goal is to establish *collaboration*—the interaction of people with similar and sometimes dissimilar interests, to jointly create and innovate—as the organizational vehicle for generating emergent solutions to product development problems. To be effective at feature-team, product-team, and enterprise levels, collaboration must be addressed in terms of interpersonal, cultural, and structural relationships.

The fourth goal of *Adaptive Software Development* is to provide a path for organizations needing to use an adaptive approach on larger projects. Because Rapid Application Development (RAD) approaches had a reputation of eschewing all discipline and rigor, they were relegated by many developers for use only on noncritical *toy* projects. *Real* projects require rigor and discipline; RAD was for playing. So, the last chapters of this book show how adaptive development works in real-life situations in which uncertainty and complexity create the need for an approach that is both adaptable *and* scalable. A major component of meeting this challenge is the ability to move from a workflow, process-oriented development life cycle to one based on work*state* and information.

The idea for this book began as one about RAD projects, written to answer such questions as, "Why does RAD work?" and "How does what works scale up?" I asked these questions because my experience and that of colleagues like Sam Bayer in implementing individual RAD projects was highly successful. I consulted with numerous large software companies that used RAD techniques on very large projects. I also worked with several IT groups that could never seem to leverage success on individual projects into larger successes. I tried to look not only at why certain practices worked, but also at the circumstances in which they were applicable.

Circumstances are defined by culture. The culture of Command-Control management has become outdated, in part because of wider cultural trends toward flexible management styles, wider participation in decision-making, and empowerment, but also because of a single pragmatic fact: Command–Control management cannot process knowledge and information quickly enough in the new economy. So, the last goal of this book is to offer a new, adaptive management style, which I label Leadership–Collaboration, to replace Command–Control. The ability to adapt and move quickly requires that "leadership" replace "command" and "collaboration" replace "control."

Our business culture and styles of management have been built around an optimizing mindset that values stability and predictability. Our management tools (derived from this belief) work—or seem to work. Consequently, we fail to understand the sharp disparity between orderly and complex systems. Orderly systems can be extremely *complicated*, but complicated and complex describe different *classes* of problems. Complicated problems yield to optimizing techniques; complex problems do not. Tools that work in orderly realms are actually antithetical to complex situations. The disparity between orderly and complex problems requires not only new tools, but a very different mindset—a significant transition for many organizations.

In an optimizing culture, increased rigor (process improvement) and stabilization are the end goal. Optimizing cultures tend to see the world as black or white, with little room for gray. If it is not rigorous, it must be chaotic (or *immature,* to use the Software Engineering Institute's parlance).

An adaptive culture recognizes gray. Researchers have coined a phrase for this turbulent gray area between order and chaos—*the edge of chaos.* It is in this variable, messy, exciting polyglot that emergence happens. In the field of evolutionary biology, for example, scientists postulate that major evolutionary advances occur in this complex region at the edge of chaos.

In an adaptive culture, the goal of rigor is to maintain balance on this edge, providing just enough stabilizing force to keep away from chaos—but no more. Adaptive organizations understand the need for *just enough rigor.* This balance does not come from a compromise of principles, but from an understanding of how the forces of optimization draw down the very energy sources that are needed to fuel emergence. Too little rigor yields chaos. Too much stifles emergence and innovation.

So, while one goal of this book is to offer emergent order as an alternative to a belief in and a dependence on imposed order, emergent order is not a complete replacement but the basis for a new, additional set of tools for managing complexity. I believe the difference in viewing optimization as a *balancing force* rather than as a goal in itself provides a significant shift in perspective.

For me, this book was a journey—one that goes on. I invite you to join and hope it engages you as it did me.

September 1999   J.A.H.
Salt Lake City, Utah

Introduction

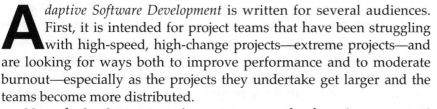

Adaptive Software Development is written for several audiences. First, it is intended for project teams that have been struggling with high-speed, high-change projects—extreme projects—and are looking for ways both to improve performance and to moderate burnout—especially as the projects they undertake get larger and the teams become more distributed.

Next, the book is written for project teams that have been assigned a high-speed, high-change project to support a critical new business initiative. These team members know their standard approach probably won't work and need a better alternative that will enable them to deliver in a culture geared to more sedate projects, often in an environment that actually is hostile to the practices required to succeed.

The third audience consists of project teams that need to accelerate schedules for small to medium-size projects. Many of the techniques described in the book were derived from this type of project. Although the limitation to scaling typical RAD techniques up to work on larger projects is change itself, adaptive development addresses this issue.

A final audience includes people who need ammunition to fend off requests for high-speed, high-change projects. Even with the best practices and effective management, extreme projects are risky. Clarifying that risk, and learning when to avoid it, is important.

Reviewing an early version of this book, columnist Warren Keuffel described how the approach differs from more procedural approaches:

> *RADical Software Development . . . is a framework within which the intelligent project manager is expected to fit the practices that have been proven to work.* —W. Keuffel [1995], p. 37.

Adaptive Software Development does not provide a set of prescriptive rules or tasks, but a framework of concepts, practices, and guidelines.

The book is divided into three parts. Part 1 consists of Chapters 1 and 2. Chapter 1 provides background material and introduces the three major models: the Adaptive Conceptual Model, the Adaptive Development Model, and the Adaptive Management Model. Chapter 2 delves into the concepts of complex adaptive systems (CAS). It then introduces the Adaptive Development Life Cycle and explains how that life cycle approach incorporates the concepts of CAS.

Part 2, Chapters 3 through 6, explains the components of the development life cycle—speculating on direction, adaptive cycle planning (detailed speculating), collaboration, and learning. The chapters of Part 2 introduce additional aspects of complexity, and then propose practical techniques based on those concepts. The chapters focus on accelerated delivery from single work groups or feature teams.

Part 3, Chapters 7 through 11, describes the adaptive management culture and practices that I have grouped under the banner of Leadership–Collaboration management. Chapter 7 provides an overview of the Leadership–Collaboration Management Model and the rationale for its use. Then, Chapters 8 through 10 explore components of the model. These chapters emphasize problems of and solutions for scaling adaptive development to larger projects. They also focus on collaboration as it embodies the concepts of emergence across multiple groups, and on the cultural and structural aspects of collaboration. Chapter 11 covers project management topics, including an explanation of a project management framework, or life cycle, and time-boxing. Project management provides a boundary of imposed order within which emergent order can flourish. The chapter treats the topics of risk assessment and other practices contributing to successful adaptive projects.

The final chapter of the book, Chapter 12, provides some parting thoughts—about dawdling, Marshall McLuhan's technology views, assessing organizational growth, and operating in thin air. Although seemingly unrelated, these topics review the message of the book in terms of a discussion for applying adaptive systems concepts.

The technology community drives unrelenting change—change in communications, in business practices, and even in business strategy. But somehow, while admonishing our businesses to adopt technology more rapidly, we, as the purveyors of technology, have failed to anticipate the impact that the speed and turbulence we have created has had on our own practice of management. *Adaptive Software Development* is ultimately about rethinking how to manage in the turbulent times we have brought upon ourselves.

Adaptive Software Development

A COLLABORATIVE APPROACH TO MANAGING COMPLEX SYSTEMS

Part 1

Part I

CHAPTER 1
Software Ascents

High on the north ridge of Mount Jefferson in the Oregon Cascades, near the top of a 25-foot ice wall—a left cramponed foot with two small front-points embedded in the ice, an ice ax pick buried a quarter-inch deep in the ice overhead, right foot scrabbling for purchase on a rock nubbin, and rear extremity hanging 800 feet above the Jefferson glacier—I asked myself, "Did I pick the wrong mountain to climb?"

There are all kinds of climbing environments, ranging from the afternoon recreational outing to those stretching the limits of human capability. Software projects are the same—from those any reasonably skilled team can complete to those only a few world-class organizations should attempt. This book describes a particular type of climbing and focuses on a particular kind of software development. It treats projects in a specific quadrant, projects that stretch the limits of achievement, demanding both high speed and high change.

Organizations get into trouble with the high-speed, high-change projects that are typical of e-business and e-commerce initiatives. Like climbers inadvertently picking a mountain beyond their climbing skills, many organizations don't know how to do these projects, and even worse, they don't know they don't know. They think the solution is simply to do more of what they have been doing.

Even companies completing projects in this quadrant routinely do so as the project team crumbles into exhaustion and stupor. Like climbers of K2 or Mount Everest, they stumble from the summit back into base camp elated at victory, but utterly wasted in body and spirit. They never want to see another mountain! Extreme sports, of which high-altitude mountaineering or solo rock climbing are examples, are defined by the risk that a single misstep will be fatal. This book targets *extreme* software projects in which the margin for error is equally small. It is intended to increase not just the staff's odds of surviving but also the chance of its emerging with reasonable mental health!

Success on extreme projects is measured in three equally important ways. First, the product gets shipped. Second, the product approaches its mission profile, a prioritized combination of scope, schedule, resource, and defect level. And third, the project team is healthy at the end.

Extreme projects are profoundly different from those less buffeted by speed and uncertainty. They require more than a new tool or technique, more than a revised software life cycle, and more than participative management. For them, success has a unique meaning:

> *Success* **is rooted in challenging our most fundamental assumptions about software development, management, and even the foundations of science.**

To understand "success" in the context of this definition, let's take a detour back into software development history.

A Historical Perspective

In the beginning, there were no methods and the world was good. Then came religion, and the great leaders were anointed by the multitude—DeMarco, Codd, Cox, Yourdon, Orr, Parnas, Booch, the list goes on. Thus began the age of methods, led by the gurus, and the world went from a state of "adhocracy" to a state of bureaucracy (see Fig. 1.1), from individuals doing their own thing, to everyone doing one thing, which was not at all what the gurus had intended, by the way. While many projects benefited from these methods, many more floundered in a sea of forms, diagrams, and repository entries. As my colleague Lynne Nix has observed:

> *An* ad hoc *approach gives the analysts and programmers an excuse not to think, while the bureaucratic approach gives managers an excuse not to think.*
>
> —L. Nix, private communication, 1996.

Figure 1.1: Bureaucracy versus "Adhocracy."

The age of bureaucratic software development was good. It was an idyllic time when mainframes ruled, clients weren't too demanding, COBOL was king, and Information Engineering was the way. Colleague Ken Orr refers to this as the era of Monumental Software Development. In more recent years, recalcitrant youngsters with powerful PCs, C++, and Visual Basic arrived at a whole new development style—one that Orr calls Accidental Software Development. These contrasting, fundamentally different styles illustrate a dichotomy in today's software development community.

Monumental Software Development

Monumental Software Development grew out of the Structured Revolution of the late seventies in which models were characterized by data flow diagrams and entity relationship diagrams. Monumental development epitomizes the saying, "Anything worth doing is worth over-doing." It is

- top-down, long-term: Start at the top of the organization; translate business needs into detailed models, particularly data models; implement those data models in databases; and then build applications. Take a year or more for planning, several years to build the databases, and then several more years to build the applications.

- analytical (engineering): Work hard and work right to yield right software—the first time.

At the pinnacle of this approach were the Monumental methodologies of the 1980's in which fourteen volumes of detailed tasks, documents, and forms defined every aspect of development. The practice led to development of the silver-bullet solution called computer-aided software engineering (CASE). In the 1990's, the mantle of Monumental development was taken up by the Software Engineering Institute (SEI) and given a new name—process improvement.

While many functional, effective software products have been built using Monumental techniques, they suffer from a series of shortcomings:

- Customers aren't satisfied and some are downright angry. At the end of the long development cycle, too many applications don't fit the customers' needs. Requirements are out of date at implementation.

- The techniques take too long. The pace of business change is accelerating too rapidly for Monumental delivery cycles. A client in the apparel business spends eighteen months doing requirements for a new system and finds that by the time requirements are finished, the proposed system is obsolete and most of the participating customers are in other jobs.

- The techniques fly in the face of political reality. All too often, these grand schemes fall apart in the middle. Executives who sign on as sponsors in the beginning become critics as the months go by. Monumental projects draw too heavily on the political capital that software development organizations are traditionally short on anyway.

In brief, Monumental development methods do not adapt well to rapidly changing conditions.

Accidental Software Development

Accidental Software Development, on the other hand, is characterized by *ad hoc* or no methods, and by a belief that processes only slow progress. It is

- bottom-up: Start with the immediate needs of a customer group and build an application to address these needs, giving little thought to the larger picture and to integration with other products.

- short-term: Limit development to two-to-six-month time frames. There is often an assumption that the business changes so fast that any project lasting longer than six months will produce an obsolete product. Many of these applications were originally intended to be throwaway.

- issue-driven: Base application justification on short-term, critical business issues. Without an overall plan, these issues are often highly politicized when priorities have to be set.

- hacker-driven: Code quickly, paying little attention to design. With short time frames and no long-term strategy, there seems little need to use slow, outdated software practices—particularly when the latest visual development tools are singing their siren song.

Accidental development also has a downside:

- Some customers are happy, but only for the short term.

- It causes stovepipe systems (narrow functionality, no integration across business functions) to be developed. With hot issues and short time frames, integration suffers.

- It results in highly redundant data. We might call this "Back to the Seventies." Stovepipe systems result in a fragmented approach to data. As client/server systems spring up in specific departments, data fragmentation and redundancy multiply. Keeping data synchronized is an increasingly challenging problem.

- The resulting software requires high maintenance. Applications typically have highly redundant data, and are built with a range of client/server tools, spread across multiple networks, using different data models. They will be tomorrow's maintenance nightmare.

The terms Monumental and Accidental are useful because they don't necessarily divide the world into software companies and internal information-technology or information-systems (IT/IS) organizations. There are both Monumental and Accidental groups existing concurrently in both types of organizations. Because of the problems of Monumental development, many companies now have warring Monumental and Accidental groups. The mainframe Monumentals perceive Accidentals as unprofessional, toy-system builders while the Accidentals perceive the Monumentals as old-fogy has-beens. The real losers in this battle are the customers who need software to run their businesses.

Obviously, Monumental and Accidental development depict extremes, with most organizations operating with aspects of both. In an era of data warehouses, client/server systems, object technology, Rapid Application Development (RAD), Business Process Reengineering (BPR), the Internet, intranets, and Java, both approaches seem somehow inadequate. Success is more than just a matter of compromising between these two views; it requires a new conceptual foundation.

Because they are staffed by both Monumental and Accidental proponents, many organizations view the world as bipolar. Existing within one organization, Accidental groups believe that creativity and innovation result when structure is absolutely minimal, while Monumental groups believe in imposed order. By force of process, planning, determinism, and control, Monumental advocates equate increasing order and structure with improving results.

"[N]atural selection is important, but it has not labored alone to craft the fine architectures of the biosphere, from cell to organism to ecosystem. Another source—self-organization—is the root source of order."
—S. Kauffman [1995], p. vii.

But there is a third alternative, which is not a compromise but is a truly different position on the landscape. Between chaos and orderliness lies the realm of the complex—a transition zone where imposed order is replaced by *emergent* order, where neither Monumental nor Accidental approaches have sufficient diversity to succeed. Emergent order is a property of complex adaptive systems, generally associated with living entities and their relationships, whose principles help us understand fields as diverse as ecology and organizational management. These principles alter our view of how the world works and, ultimately, provide a different perspective on how organizations perform in increasingly accelerated and uncertain environments.

A Rebirth in World View

W. Brian Arthur's article in the *Harvard Business Review*, entitled "Increasing Returns and the New World of Business," culminated more than two decades of his trying to convince mainstream economists that their world view, dominated by fundamental assumptions of decreasing returns, equilibrium, and deterministic dynamics, was no longer sufficient to understand reality. The new world is one of increasing returns, instability, and an inability to determine cause and effect.

Economics based on decreasing returns has its roots in the late 1800's and is characterized by bulk processing of goods—most notably, of iron, steel, and agricultural products. The equilibrium economic theory that developed during this period emphasized that production was repetitive, orderly, and predictable—in Arthur's word, "genteel."

The world of increasing returns is much different. Microsoft, Sun, Oracle, Cisco, and other high-technology companies are powered by knowledge rather than by bulk-processing skills. Like the PC-operating-systems market dominated first by DOS and then Windows, these increasing-returns markets are characterized by

> [I]nstability (the market tilts to favor a product that gets ahead), multiple potential outcomes (under different events in history, different operating systems could have won), unpredictability, the ability to lock in a market, the possible predominance of an inferior product, and fat profits for the winner.
> —W. Arthur [1996], p. 102.

These two worlds of bulk-processing and knowledge-based economics impose very different demands on management and require very different cultures. Bulk processing's repetitive nature leads to a static environment of planning and control in which optimizing techniques designed for efficiency are paramount. Knowledge-based companies are dynamic, always searching for the next big winner. For them, an emphasis on optimization is anathema. In these companies, flexibility and speed are everything.

In knowledge-driven, high-tech industries, the competitive game changes frequently and dramatically. Optimization does not make sense in this environment. "Adaptation is what drives increasing-returns business, not optimization" (Arthur96, p. 107). Adaptation is

"Our understanding of how markets and businesses operate was passed down to us more than a century ago by a handful of European economists. . . . It is an understanding based squarely upon the assumption of diminishing returns. . . . Increasing returns generate not equilibrium but instability. . . . The two worlds . . . differ in behavior, style, and culture. They call for different management techniques, strategies, and codes of government regulation. They call for different understandings."
—W. Arthur [1996], pp. 100, 101.

an organism's or an organization's ability to alter its internal rules of operation in response to external stimuli.

Complex Adaptive Systems

Arthur's work extends into the management arena concepts that have shaken the foundations of science during the last ten to fifteen years—complex adaptive systems (CAS) theory.

Complex adaptive systems theory arose early in this century's exploration of physics. As smaller and smaller atomic particles were discovered, Newtonian physics, based on predictable certainty of determinism (cause and effect) and reductionism (reducing objects to their component pieces), gave way to quantum physics, a science that was based on probabilities. The realization that the old world of physics was crumbling, that the very foundations were shifting, was traumatic.

From physics to biology to chemistry to evolution, complex adaptive systems theory began to help explain occurrences in the real world that the linear approximations of the old science could not. Traditional management theory, derived from determinism, is mechanistic scientific management, in which organizations are viewed as composed of parts. Scientific management attempted to equate the performance of people with that of factory machines—each was a different type of part. And while more participative styles have influenced recent management literature, these styles have not had a scientific foundation of the stature of those laid out by Newton or Darwin. Business management practices today seem particularly wedded to Darwin's theories of natural selection and survival of the fittest.

What if Darwin was wrong? Even Darwin himself was uncertain about natural selection's ability to account for evolution's variety. His evolutionary theories of random mutation and survival of the fittest are difficult to corroborate using mathematics. The problem with survival as the primary evolutionary mechanism is that even given geologic eons, the mathematics of random mutation does not provide sufficient time to produce today's complex biological diversity. As some critics of Darwin have quipped, it has the same probability a tornado sweeping through a junkyard has of assembling a Boeing 747. There is a growing contingent of evolutionary biologists, represented by the work of Stuart Kauffman of the Santa Fe Institute, who insist on *arrival*

"Ever since Darwin, there has been a tension between two schools of thought about how to carve up the biological world. The strict adaptationists believe that natural selection alone is powerful enough to explain most of the order, while a school called the structuralists insists that something extra—laws of self-organization—is required."
—G. Johnson [1995], p. 235.

of the fittest, not *survival of the fittest*, as the driving force behind evolution. Moreover, where random mutation provides the grist for the survival mill, self-organization does the same for arrival.

Self-organization is the tendency for living things to work together for some common purpose in the absence of some central organizing force. For example, single-cell eukaryotes band together and begin to cooperate, forming a new multicellular organism. Once higher-level organisms emerge, natural selection is the mechanism for further refinement. The new evolutionary theory postulates creating significant new life forms through self-organization and then refinement of those forms by natural selection.

Complex adaptive systems, self-organization, emergence, and nondeterminism are all scientific, descriptive, but somewhat sterile terms. What they mean is that organizations—project teams—are not machines, but *living organisms*. Companies have long been viewed as things, as machines to produce profits by grinding the cogs (people) together in certain patterns. However, companies are composed of people, not people as human resources to be reengineered, but people as a community organized for a certain purpose. How have we come to view an aggregate of living beings as a dead, mechanistic, linear machine? How has this view inflicted sickness on the organisms?

Would a lion killing a wildebeest and then dying from eating the flesh provide a viable model for a natural organism? The lion succeeded by killing the prey (shipping the product), but died in the process (burned out, left the company)—not a viable strategy for a lion or a project team.

Approaching software development as an adaptive process, and viewing the project team itself as a living organism rather than as an impersonal machine, provides a better model for managing extreme software projects. Such an approach produces better products more quickly, and at the same time fosters healthier organisms ready to tackle the next project.

A New World View of Software Development

In the world of software development, it seems the Software Engineering Institute's Capability Maturity Model (CMM) is the Holy Grail. Even nonbelievers wonder if there is some magic they are missing. But on what basis has the CMM taken on the mantle of The One True Reli-

"Like all organisms, the living company exists primarily for its own survival and improvement: to fulfill its potential and to become as great as it can be. It does not exist solely to provide customers with goods, or to return investment to shareholders, any more than you, the reader, exist solely for the sake of your job or your career."
—A. De Geus [1997], p. 11.

gion? What emboldens CMM supporters to apply the model to all software development?

If I were stepping into a space shuttle about to be rocketed into space, the fact that Lockheed Martin's Orbiter Avionics software team was rated CMM level 5 would calm at least some of my anxiety. One trivial defect in 420,000 lines of code is a Monumental achievement.

However, if I were a manager at Netscape or Microsoft, and the Orbiter Avionics group announced it was going into Web-browser competition, I would be amused. A level-5 CMM group would not survive long in the rough and tumble world of Internet speed. In this world, optimization is necessary, but insufficient for success.

In Arthur's terminology, the Orbiter Avionics group, even though it produces a complicated product, operates in a decreasing-returns environment, whereas Microsoft and Netscape operate in one of increasing returns and complexity. In complex environments,

> *Adaptation* **is significantly more important than optimization.**

In this context, adaptation means more than the ability to respond to environmental stimuli. Adaptation includes the ability to utilize emergent order to alter actions that are essential if an organization is to survive and thrive in complex social and economic ecosystems. It includes the ability to make local alterations rather than depending on centralized, slow acting, control processes. Adaptation trades efficiency for speed and flexibility. The CMM is a wonderful optimization model; it is not a wonderful adaptation model. Optimization works in a complicated world; adaptation works in a complex one.

The unleashing of emergent order, as contrasted with imposed order, is the key to adaptation. Emergence is a property of complex adaptive systems that creates some greater property of the whole (system behavior) from the interaction of the parts (self-organizing agent behavior). This emergent system behavior cannot be fully explained from measured behaviors of the agents. Emergent results cannot be predicted in the normal sense of cause-and-effect relationships, but they can be generated by means of patterns that have previously produced similar results. Therefore, in complex environments,

> *Emergence,* **characterized as** *arrival* **of the fittest, is significantly more important than** *survival* **of the fittest.**

As software developers, we are sometimes so enamored with the competitive survival model, we miss opportunities to arrive as part of a team at a much higher peak than we could scale alone. Arrival of the fittest is the engine that powers adaptation.

"Only at the edge of chaos can complex systems flourish." —M. Crichton [1995], p. 2.

To manage speed and change in software development, we must face a dichotomy of old and new, of optimization versus adaptation. This book is about producing software in a world of increasing returns, one in which competition is unpredictable, nonlinear, and fast. Competition in this world—in the Internet industry, for example—is like the start of a motocross race. Fifty high-powered motorcycles at the starting line, and only enough room for three or four at the first turn fifty yards away. Traditional practices put your bike at the back of the pack, or out of the race altogether.

The theory of arrival of the fittest points us in the direction of better cooperation and creative collaboration. *Co-opetition*—the balancing of competition and cooperation—in today's businesses functions much the same as cells cooperating to form higher-level organisms, which in turn are better competitors in the environment. The myriad of mergers, joint ventures, and collaboration in the high-tech marketplace is indicative of the benefits of co-opetition.

The Challenge of Understanding

As CAS theory affords new insights into the natural world, it also provides new metaphors and new strategies for management. Managers steeped in optimization and survival strategies need to consider how to better utilize those of *adaptation* and *arrival*. The challenge of learning to apply fundamentally different approaches is formidable. So, whether we come from the Accidental or the Monumental camp,

> **We must challenge our most fundamental assumptions about software development.**

Dilbert, perhaps the foremost management guru of our time, puts the problem most succinctly:

> *People are idiots.* —S. Adams [1996], p. 2.

What I value most about Dilbert is the lesson underlying the humor. Scott Adams, Dilbert's creator, shows how we are idiots in some way nearly every day. If we apply the Dilbert humor to managing in an extreme environment, and ask ourselves, "What is the greatest risk we face in developing software?" the answer could be stated,

> **The greatest risk we face in software development is that of overestimating our own knowledge.**

This statement gets me off the hook with clients. Somehow, starting off engagements with new clients by telling them they are idiots doesn't go over very well. Scott Adams might be able to do it, but most of the rest of us can't. At the core of our ability to succeed in extreme environments is the admission that we don't know it all, that we need to enrich our understanding through better attention to how we explore our world, to how we collaborate, to how we learn—in essence, to how we adapt as living systems.

Components of Adaptive Software Development

This book offers an approach, Adaptive Software Development (ASD), as a framework from which to address the issues of high-speed, high-change software projects. The approach is based on my years of practical experience with traditional software development methodologies, on my consulting and writing about RAD techniques, and on my work with high-technology companies creating better ways to manage product development. From a conceptual perspective, ASD is grounded in the emerging science of complex adaptive systems theory, which helps explain why many of the experiential projects have worked.

There are three interwoven components in Adaptive Software Development:

- the Adaptive Conceptual Model
- the Adaptive Development Model
- the Adaptive (Leadership–Collaboration) Management Model

Without practical techniques, conceptual ideas remain in the clouds. Conversely, techniques without a theoretical base are reduced to a series of steps executed by rote. Concepts and practices reinforce and strengthen each other. Extreme projects don't succumb to rote practices; they require judgment based on a firm, conceptual foundation.

Extreme environments move quickly, demanding fast learning among project members and often forcing people to abandon preconceived assumptions. Fast learning requires iteration—try, review, repeat. Accelerated schedules demand a high degree of concurrency, with developers working on many components at the same time. Taken together, iteration and concurrency generate high levels of change, especially as project size escalates.

The Adaptive Conceptual Model introduces the new science of complex adaptive systems as the conceptual foundation for both development and management. The Adaptive Development Model focuses on iterative phases of development and on work-group-level practices to increase speed and flexibility. The Adaptive (Leadership–Collaboration) Management Model focuses on forging an adaptive culture and identifying adaptive practices, particularly those involving distributed work groups and dealing with high change, collaboration, and management of results.

The Adaptive Conceptual Model

The concepts underlying Adaptive Software Development are derived from complex adaptive systems theory, which a small but growing group of scientists believes is revolutionizing our understanding of physics, biology, evolution, and economics. The key components of CAS theory are agents, environments, and emergent outcomes. A complex adaptive system exists in an ecosystem, which is created by the presence and interaction of those agents—in economic terms, a market.

A complex adaptive system is an ensemble of independent agents

- who interact to create an ecosystem,

- whose interaction is defined by the exchange of information,

- whose individual actions are based on some system of internal rules,

- who self-organize in nonlinear ways to produce emergent results,

- who exhibit characteristics of both order and chaos, and

- who evolve over time.

"[O]ur organizations work the way they work, ultimately, because of how we think and how we interact. Only by changing how we think can we change deeply embedded policies and practices."
—P. Senge [1990], p. xiv.

Understanding these characteristics, and their applicability to managing software development, is a central theme of this book. Another is to present and explain the concepts of CAS as a framework for software development, providing a balance between theory and actionable practices. By viewing software product development groups as complex living systems, we can use our understanding of adaptation as a lens through which to see, develop, and evaluate principles and practices to improve product quality and time-to-market.

We also must identify fundamental concepts and principles at the heart of why people are motivated to action. With virtually hundreds of software-engineering, team-building, project-management, and quality-assurance techniques to choose from, organizations can hope to succeed only by explicitly exploring the governing principles motivating the people who must use the techniques.

Understanding techniques but not principles limits effectiveness. As a sophomore in college, I found my first engineering class to be a frightening experience. For the first exam, I dutifully studied the book, worked the homework problems, memorized the formulas, and *failed* the exam. In fact, the class average was somewhere in the 40s. We understood the practice (the formulas), but not the principles on which the practice is based. The exam problems required thinking—utilizing principles to derive the needed formulas, which were slightly different from the standard ones in the engineering textbook.

In exploring organizational effectiveness in his *Quality Software Management* series (Weinberg92–97), Jerry Weinberg defines key differences between behavior patterns: Steering (Pattern 3) organizations continue to use their practices in a crisis, whereas Routine (Pattern 2) organizations tend to abandon them. Steered teams believe in what they are doing. Routine teams perform because someone told them to.

One reason why steered organizations *believe* in their processes is that they understand the concepts as well as the practice. When the practice doesn't apply exactly, they have the underlying knowledge to adapt the practice to fit the situation.

The Adaptive Development Model

In the early 1990's, Rapid Application Development emerged as a development method in reaction to the need for shorter software delivery cycles and an antidote to the plodding, excruciatingly detailed Monumental methods. It recognized the business need for speed, which customers associated with excellence, but it also became associated with hacking, lack of good engineering practices, and small projects—in other words, with Accidental development.

RADical Software Development, my own contribution to RAD approaches, is a synthesis of RAD and best-practice software engineering techniques. RADical reflects a combination of two words: RAD, for Rapid Application Development; and radical, for a different model of software development.

The push for speed and the industry's foray into RAD and RADical techniques had unforeseen consequences. In pushing for speed, we learned more about collaborating, dealing with uncertainty, iterative learning, working with customers, and synchronizing concurrent development efforts.

In working with high-tech software developers who used RAD-like techniques, I became aware that RAD, and even RADical, was not enough. What was needed was a more fundamental conceptual base and practices that could scale up to larger, more complex projects. From these beginnings, RAD became RADical, and then that became Adaptive Software Development, shown graphically as

RAD ⇨ RADical ⇨ Adaptive

Figure 1.2, below, shows components of the Adaptive Development Model. While this model is iterative, similar to a spiral or evolutionary model, the nature of the iterations is different. At its core, adaptive development reflects the developer's mental switch from an inexorable belief in imposed order—a linear, deterministic view of the world—to an acceptance of emergent order as a source of solutions in complex realms. The components of the life cycle are structured to convert this concept into actionable practices.

"Shortening the development cycle is a tool that no company can afford to ignore if it wants to remain viable in the 1990s. . . . Development time . . . is also in a sense a surrogate for effectiveness of the product development process."
—P. Smith and D. Reinertsen [1997], pp. 3, 10.

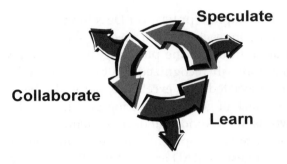

Figure 1.2: Adaptive Development Model Components.

Thriving on Speed and Change

Adaptive development focuses on products and projects with three characteristics—high speed, high change, and high uncertainty. As Fig. 1.3 shows, different approaches are appropriate for different rates of speed and change.

	Low Speed	**High Speed**
Low Change	Waterfall	RAD
High Change	Evolutionary	Adaptive

Figure 1.3: Life Cycle Matrix.

The waterfall approach, which proceeds in a serial fashion from analysis to design to coding and testing, has fallen into disfavor in recent years. However, there are still situations in which requirements are well-known and the risk of change is small. In those cases, the waterfall approach is preferable. Tom Gilb used the term "evolutionary" in the mid-1980's to describe an iterative approach with short, well-planned development cycles. Barry Boehm's spiral model is similar in its iterative cycles but explicitly incorporates the concept of risk-driven cycles.

RAD practices are associated with speed, but as project size increases, high-change rates and larger team sizes have limited the effectiveness of using RAD approaches. The typical techniques employed on a RAD project are usually geared to low-to-moderate levels of change. While the conclusion that RAD approaches are low change may seem odd, with the typical smaller size of a RAD project, the total volume of changes is still relatively small.

Evolutionary techniques have been successfully used on very large projects to help manage requirements changes. Evolutionary techniques by themselves allow product development to react to change, but often at the expense of speed. Even RAD and evolutionary techniques used together are not enough to handle accelerating uncertainty, speed, and change.

There are a couple of distinctions regarding complexity, particularly as it relates to project size. First, increasing size seemingly would lead to increasing complexity, but that is not necessarily so. Large size does complicate a project, but a large, slow, low-change, relatively predictable project is not a complex one. Increasing complication can be solved by increasing optimization; increasing complexity cannot.

Second, the primary driver of high change is not size but uncertainty about factors such as requirements. Uncertainty is only a possibility, which, when manifested, results in an actual change. While larger size results in higher levels of internal change (requiring good configuration management, for example), it is the increase in external uncertainty (resulting from factors such as technology and competition) that drives projects into the high-change category.

Many of the practices of ASD are oriented to high speed—JAD, iterative cycles, time-boxing, risk-driven planning, and concurrency—but in real-world projects, they bump up against an important fact:

High speed is easy; high change is the real challenge.

My colleague Sam Bayer and I have been involved in dozens of successful RAD projects that have been completed on schedule and that have satisfied clients. But success on individual projects has not always resulted in longer-term organizational change. It finally occurred to me that one real problem is that many of the companies we work with are in decreasing-returns industries, and *do not need to go faster!* Of course, nearly everyone thinks faster, but when survival is dependent on being fast—for example, in the software markets of Microsoft, Netscape, and Oracle—the word takes on a completely different meaning. Significant change, such as greatly accelerating development, must be driven by a critical motivating factor—the marketplace. In many companies, the need for speed is just not there—and possibly doesn't need to be there.

The need to manage high change pushes us into the realm of adaptation more forcefully, and dictates a revised management model.

The Adaptive (Leadership–Collaboration) Management Model

"Usually, everybody's zooming too fast to notice the décor. There are impromptu meetings aimed at getting quick decisions. They call it 'surround-sound' management: an issue comes up and somebody instantly calls a meeting of key people. They swarm on the problem for 20 minutes, come to a conclusion, and separate."
—R. Hof [1997], p. 48.

All of us fall into the trap of using clichéd phrases about the constancy and pervasiveness of change—yet our conceptual framework and management practices still view change as an exception. Most of our concepts, practices, and tools are geared to environments in which equilibrium is the normal condition and changes are the exception. Extreme, high-speed, high-change environments are just the opposite—change is the norm.

> **In extreme environments, *equilibrium* is the exception condition.**

In such circumstances, change management is not an add-on procedure to a linear process temporarily in disequilibrium—change management is the heart of the development process itself.

Adaptation is a pattern of continuous change that exists in contrast to the periodic discrete change process found in many organizations. The difference between beginner and expert skiers provides a vivid analogy. The beginner traverses the slope until he or she encounters the trees at the edge. For skiers, trees provide a significant incentive to change (turn). However, the expert skier is always changing, always turning, always on edge, always adapting to the challenge down the hill. The beginning skier is more like a traditional organization, utilizing existing practices until the threat (or actuality) of encountering trees is so great that he or she (or it) has to change. Adaptive organizations, like advanced skiers, treat continuous change as the norm—and their practices reflect that actuality.

Adaptive change is graceful because it flows from all levels of an organization. Historically, however, change has been imposed in a top-down manner. From Newtonian determinism came Frederick Winslow Taylor's "Principles of Scientific Management" (Taylor16), first published in 1916, which spawned our almost slavish focus on process and workflow. Early science and early warfare laid the foundation for a command-and-control management philosophy in which it was thought that since the manager *knows* the objective, he or she can *command* the troops to conquer the objective—that is, once the command is given, the manager monitors progress and thereby *controls* the outcome. This approach worked well as long as the objective to be conquered did not move around much, and as long as the organization existed in the more predictable world of decreasing returns.

Adaptation cannot be commanded, it must be nurtured.

Adaptation depends on leadership and collaboration rather than on command and control.

The focus of collaboration in the Adaptive Development Model is on the work group and interpersonal relationships so as to create emergent order locally. In the Adaptive (Leadership–Collaboration) Management Model, the focus is on the cultural and structural aspects of collaboration necessary to create emergent order more globally—to scale adaptive development up to larger complex projects.

The structure of an organization's collaborative network has significant impact on its ability to produce emergent results and ultimately on its very ability to adapt.

Leadership focuses on creating the cultural environment in which adaptation and collaboration can thrive, and on creating a collaborative structure in which multiple groups can interact effectively.

The first major strategy for managing high change through creation of a collaborative structure requires deployment of methods and tools that apply increasing rigor to the results, that is, to the work*state* rather than to the workflow.

In product development, indeed in any endeavor in which a degree of creativity and innovation are involved, we must abandon the workflow mentality and apply increasing rigor to managing the results. Even high-tech independent software vendors (ISVs) have *process envy* as they grow larger. They harbor a feeling that they need to be more process-oriented, but they really don't want the anticipated added bureaucracy. The process workflow mentality is so engrained in our business culture that we experience a latent feeling of immaturity if we are not process-oriented.

> **The second key strategy for managing high change through creation of a collaborative structure requires deployment of methods and tools that support self-organizing principles across *virtual* teams.**

Rapid Application Development works well when teams are small, dedicated, and collocated. In fact, this *teamwork* may be the single most important reason RAD projects are successful. Unfortunately, there are many projects requiring more than a half-dozen people for six months, projects in which participants are drawn from across a country and from around the globe. These "virtual" teams are separated by distance and time. RAD teams succeed because of intense interactions within a small group. Maintaining this high degree of interaction despite a team's being distributed across time and space is essential to success in complex environments.

Integrating the Models

As we've seen, the three models of Adaptive Software Development are the conceptual, development, and management models. A central theme of emergent order links these models together and separates them from the models used in traditional development.

In Fig. 1.4, three interlocking circles are used to depict both adaptive and traditional development. At the core of the traditional approach is a belief in and actions to promote imposed order—that is, the ability to predict the future and make it happen. At the core of the adaptive approach is the more powerful, although less-well-understood property of *emergence*, an ability to produce results through spontaneous interaction of self-organizing agents.

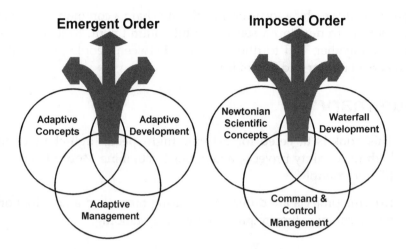

Figure 1.4: The Core Difference Between Adaptive and Traditional Models.

The Road Ahead

Software development, like mountaineering, is a dangerous activity. Software project failures are frequent and expensive. Climbing failures result in injury, and sometimes death. But we still attempt both, because they allow us to experience what is clichéd as "the thrill of victory" while "staving off the agony of defeat." Techniques don't climb mountains—people climb mountains; metaphorically, people and teams must climb software mountains.

The core of success is creative collaboration and individual responsibility. It's easy to find fault with, and thereby abdicate responsibility to, the ubiquitous "they"—management, customers, users, other teams, anyone but ourselves. As a member of a climbing team scaling the 1,800-foot north ridge of Mount Stuart in the Washington Cascade mountains, I experienced firsthand that a climber has nowhere to delegate responsibility for success, or failure, but to himself and his climbing team. The responsibility clearly must be on *me and us.*

The target environment of this book is like a fierce mountain with steep sides, avalanches, falling rocks, hidden crevasses, and quick-rising storms—a high-speed, high-change mountain. Getting to the top, sharing commitment, learning as the climb progresses, balancing safety and speed—these are all part of the journey. Adaptive Software Development is driven by a focus on people—to share a common mission, to collaborate, to develop a sense of joint commitment, to learn on the journey together, and to find a balance between people and product and between rigor and flexibility.

Summary

➤ This book is targeted for extreme high-speed, high-change, and high-uncertainty projects, analogous to climbing Mount Everest or K2, for example.

➤ Extreme projects are different, requiring more than a new tool or a new technique; they require a new understanding:

> *Success is rooted in challenging our most fundamental assumptions about software development, management, and even the foundations of science.*

➤ Historically, there has existed a dichotomy between Accidental and Monumental development methods. The solution to managing extreme projects is not a compromise between the two, but a new model entirely.

➤ W. Brian Arthur explains, from an economist's perspective, the differences between companies in increasing-returns versus decreasing-returns markets.

➤ Thriving in decreasing-returns markets is based on Newtonian determinism and optimization.

➤ Thriving in increasing-returns markets is based on the emerging science of complex adaptive systems and adaptation.

➤ A complex adaptive system is an ensemble of independent agents

- who interact to create an ecosystem,

- whose interaction is defined by the exchange of information,

- whose individual actions are based on some system of internal rules,

- who self-organize in nonlinear ways to produce emergent results,

- who exhibit characteristics of both order and chaos, and

- who evolve over time.

➤ From this new science come the two most fundamental concepts for managing in complex, increasing-returns markets:

> *Adaptation is significantly more important than optimization.*

> *Emergence,* arrival *of the fittest, is significantly more important than* survival *of the fittest.*

➤ Adaptation is the ability to utilize emergent order to alter actions essential to surviving and thriving in complex social and economic ecosystems.

➤ Adaptation involves facing our own fallibility:

> *The greatest risk we face in software development projects is that of overestimating our own knowledge.*

➤ Adaptive Software Development is a framework of concepts and practices targeted for extreme projects. It incorporates experience in RAD, working with commercial software vendors, and the fundamental view of software development groups as complex adaptive systems.

➤ Adaptive Software Development consists of three models: an Adaptive Conceptual Model (complex adaptive systems); an Adaptive Development Model (speculate, collaborate, learn); and an Adaptive Management Model (leadership and collaboration).

➤ The conceptual model formulates the mental model, the mindset, needed to effectively utilize the development and management models.

➤ The development model focuses on high speed through the use of iteration, concurrency, feedback, and collaboration.

➤ High speed is easy; high change is the real challenge.

➤ The world of adaptation depends on leadership and collaboration rather than on command and control.

➤ Success with high change involves

*Deployment of methods and tools that apply increasing rigor to the results, that is, to the work*state *rather than to the workflow.*

Deployment of methods and tools that support self-organizing principles across virtual *teams.*

➤ The structure of an organization's collaborative network has significant impact on its ability to produce emergent results and ultimately on its very ability to adapt.

CHAPTER 2
Thriving at the Edge of Chaos

Cecret Lake lies in a shallow basin beneath Devil's Castle and the 11,051-foot Sugarloaf Peak above Alta, Utah. Renowned for skiing, Alta in the summertime displays a profusion of alpine wildflowers—mountain bluebells, Indian paintbrush, lupine, columbine, and larkspur. The trail from the campground parking lot to Cecret Lake rambles three quarters of a mile across melt-water streams and past rocky ridges, rising 420 feet before it reaches the shore of this small alpine lake. On a pretty Sunday in July or August, the trail is awash with humanity absorbing the mountain splendor.

Venturing up another 800 feet, across a half-mile of steep talus slopes, a hiker can reach the low point in the ridge between Devil's Castle and Sugarloaf Peak. By pushing on another quarter mile and 500 feet higher, the hiker arrives at the panoramic summit of Sugarloaf. Those hikers choosing to set off in the other direction, half a mile along a precarious ridge, will arrive at the highest spike of 10,920-foot Devil's Castle—not a trip for the timid or height-challenged hiker, but one that can be safely undertaken by a fit day-hiker.

Down from Alta in Little Cottonwood Canyon is the most popular rock-climbing destination in the area—an array of vaulting granite spires called Gate Buttress. One of the most spectacular climbs is the moderately technical, 600-plus-foot ascent called the Thumb. After a steep scramble up a talus slope, the climber must struggle through an underbrush-clogged dihedral before angling off up a small crack system. Continuing on around a corner named Indecent Exposure to a traversing ledge (Lunch Ledge), the climber can watch cars snake up the canyon hundreds of feet below. The next pitch involves an off-width crack—too wide for finger and hand jams, too narrow for one's body to wedge into—where climbers mimic inchworms in their ascent. Near the top, a single step across a gaping chasm puts the climbers on top of the thumb—a broad spine with a majestic view below. To reach this second venue requires that a climber have considerably more experience than a hiker at Devil's Castle or Sugarloaf Peak, but it is not the most difficult challenge to be found near Alta.

Still more difficult is the terrain south of Salt Lake City, around the point of the mountain and split by Highway 92—American Fork Canyon. Known as AF to the locals, it is home to some of the most difficult rock climbs in the world. Along Highway 92 just past Timpanogos Cave National Monument and less than 200 yards from the road lies Hell Cave, where climbers traverse the horizontal ceiling of the limestone cavern like mutants held in place by some unknown force of nature. Certainly no ordinary human could do what they do.

At a 5.14a difficulty rating, three grades short of the hardest climbs in the world, Dead Souls is a world-class climb in Hell Cave that only a select few have mastered. The route is only about 35-feet long, but it is extremely overhanging, and most moves tax even the best of the best. Dead Souls provides an extra degree of excitement because it is so intensely steep (like hanging across a ceiling) that a climber who slips while clipping to one of the protective bolt anchors in the rock face in all probability will fall straight to the ground.

The entire climb is excruciatingly difficult, but the most demanding section requires that 12 complicated moves be made along only 20 feet of overhanging wall. Most of the handholds are crimpers—small ledge-like holds less than the width of a fingertip—or monos—pockets in the rock large enough for only a single finger to grasp. Every move requires strength, muscle tension, accurate foot placement, and precise control of body position.

These three climbing environments can be thought of as *fitness landscapes,* to use the terminology of complex adaptive systems; that is, they are landscapes in which fitness is tested. The first, a short hike to Cecret Lake, could be successfully accomplished by tens of millions of people. The hikes above Cecret Lake would reduce the numbers of adventurers substantially, especially the scramble to the top of Devil's Castle. The Thumb vaults participants from the level of a fit day-hiker into the realm of technical climbing—ropes, belays, carabiners, tapers, hexes, and camming devices—and much greater risk. But of the hundreds of thousands of climbers in the United States, many would be capable of ascending this route. The third, Dead Souls, is an entirely different matter, however. Successful ascent of it is within the reach of probably no more than one hundred of those hundreds of thousands of climbers in the country.

Different fitness landscapes, whether they consist of mountains or software projects, require different approaches and skills. As we have seen, these landscapes can vary from the relatively benign to the complicated to the complex—for complex software projects, the landscape is analogous to Dead Souls, and the projects push teams to the edge of chaos.

In chaotic environments, success is accidental. In stable environments, success can be achieved by a wide variety of means. To manage projects at the edge, where stability blurs nearly into chaos and where, as at Dead Souls, a small miscalculation will surely mean failure, we need to fully comprehend how organizations must work to achieve success. Understanding climbers gives us a perspective on how to ascend increasingly difficult terrain. Understanding fitness landscapes (like a mountain range) provides us with a perspective on projects. Complex adaptive systems theory, based in natural and physical science and extended into organizational science, provides a key to this deeper understanding.

Fitness landscapes, whether they are in the realm of natural ecologies or market economies, are one part of the fabric of complex systems. A second essential part consists of the people (or in market-economy terms, the agents) populating the landscape. The following section treats the people component in more detail.

People as Agents

In *Performance Rock Climbing* (Goddard93), Dale Goddard and Udo Neumann provide several metaphors that are especially interesting in the context of the people aspect of complex projects. As superb climbers, Goddard and Neumann are an inspiration to all who encounter them, but their greatest contribution to their sport is in presenting a holistic approach to improving performance.

Early in their book, Goddard and Neumann introduce three characters. First is Bruno, a somewhat macho, male ex-bodybuilder with bulging arm muscles and a bull-in-the-china-shop mentality. Bruno's approach to any climb is brute strength. Second is Julia, a lithe, flexible athlete who uses trunk muscles, body position, and fluid movement to achieve the same climb Bruno has just powered up. Third is Max, who is young and full of himself, but whose compact, high-strength, sinewy muscles, and lightweight frame give him the perfect body type to be a world-class climber.

Goddard and Neumann use the three characters to illustrate that each person has a set of unique strengths to build upon, communicating clearly the message that no single strength assures success. Although Max may have a better body type for the sport, a lax mental attitude may significantly impede his progress. Julia may not have raw strength, but her agility and gymnastic training allow her to climb steep routes. To the non-climber, climbing may look like a strength sport, but it is a ballet consisting of an intricate combination of steps, movement, strength, coordination, and mental attitude. To succeed at world-class rock-climbing levels, a climber must master the ballet. Each person's ballet may have different components, but no single component is sufficient by itself.

Software development requires a similar ballet, especially as the environment becomes more complex and challenging. It requires a ballet that balances knowing and learning, process and people, concepts and practice, rigor and flexibility.

Max, Julia, and Bruno are climbers with different perspectives, abilities, and mental attitudes. Cecret Lake, the Thumb, and Dead Souls are environments that require climbers with different abilities and perspectives. Climbing is about the combination of two things—climbers and climbing environments. Software development, at its essence, is about two things also—developers and markets. It is about delivering products in environments as benign as that of Cecret Lake or as imposing and demanding as the craggy ceiling of Hell Cave.

But while the analogy between climbing and building software provides us with images that help to illustrate the concepts of fitness landscapes and agents, there is a third major part of the fabric of complex adaptive systems that is crucial to solving complex software development problems—emergence.

Emergence and the Flocking of Boids

Emergence is a property of complex adaptive systems that creates some greater property of the whole (system behavior) from the interaction of the parts (self-organizing agent behavior). Emergence is similar to innovation, problem-solving, and creativity; we have a difficult time understanding how they happen, but the results are clear. Emergence can be seen in both organizations and biological systems.

Every great athlete, at some point, experiences being "in the zone." Probably the most famous example of this is Michael Jordan. When Jordan *elevated* his game, the Chicago Bulls were nearly unbeatable. The same phenomenon can be at work in product development. For example, in *The Soul of a New Machine* (Kidder81), Tracy Kidder tells the story of a team of computer developers who were in the zone—Kidder calls it a "groove"—and who functioned as a team that elevated its game to a level few outside the team thought possible.

Although we cannot adequately explain it, we have words to describe this phenomenon—in the zone, in the groove, elevated game, creative. We know it exists, we can see it happening, we each have experienced the feeling—a ghostly presence seeming to come and go at a whim. And yet, because we cannot map the effect back to a specific cause, our belief in cause and effect often influences us to discount this phenomenon—if we cannot plan it or predict it, how can we possibly use it in the concrete world of business?

Those scientists at the forefront of the movement to understand complex systems have labeled this phenomenon "emergence." It is fundamental to the concepts of both *adaptation* and *arrival of the fittest*. One of the most difficult concepts in adaptive development to internalize is the idea that although the direct link between cause and effect may be broken, emergence is a strong, indirect link. We develop patterns of behavior (practices) that generate desired results, but we never seem to know exactly how the pattern worked. A better understanding of emergence will help us utilize these patterns even when we don't fully comprehend them.

The concept of emergence is well-illustrated by complex systems research done in the mid-1980's. Craig Reynolds created a computer simulation to capture the essence of the flocking behavior of birds (Waldrop92). Each "boid" (Reynolds' name for his simulated birds)

"[Emergence is] a global property of a complex system. . . . For example, consciousness is an emergent property of the many neurons in a human brain."
—P. Coveney and R. Highfield [1995], p. 426.

followed three simple rules of behavior. The boids' pattern was to separate, align, and cohere, as they attempt one of three behaviors:

1. They attempt to maintain a minimum separation from other boids.

2. They attempt to align velocities with other nearby boids.

3. They attempt to move toward the center of mass of nearby boids in order to cohere with them.

Reynolds' boids did not follow any rules for group behavior. There are no algorithms defining the results expected from the group, but only rules about the behavior of individual boids. Yet the complex behavior of the group, an emergent behavior not specified by the simple rules, is one of flocking. Groups of boids flow over the screen landscape. Errant boids, momentarily distracted by digital barriers, rush to catch up to the flock. The flock splits around obstacles and reforms on the other side.

Imposed order is programmed. Emergent order *happens*, not in some mysterious way, but as a result of intelligent interactions of agents striving for a better result. Emergent order results from patterns in our complex world—patterns we may not completely understand, but which we can certainly use. Most innovative software design is emergent. Complexity yields to concerted, nurtured, encouraged interaction of agents governed by simple rules. Complex rules, administered through limited interactions, yield adequate results only within simple, relatively stable, situations. Complex behavior works according to the basic equation,

Complex Behavior = Simple Rules + Rich Relationships

Artificial Life (A-Life) is one field of study within complex systems. It is devoted to creating and analyzing lifelike organisms, particularly in digital form, although sometimes in robotic form. Artificial Life and Artificial Intelligence (AI) reflect two different philosophies about creating complex behavior. AI is brain- or central-processing-unit-centered. Sensory inputs are monitored, a complex rule-driven analysis is performed, and prescribed actions are initiated. Building the rule base is the crux of the design problem.

A-Life takes a very different approach. Whereas AI is top-down and rule-based, A-Life is bottom-up and emergent. As in the flocking of boids, the behaviors of the whole are not programmed. An A-Life

"There are, then, two diametrically opposed ways of answering some of the most basic questions to be posed about organizational life. One . . . leads to ways of managing and organizing that greatly constrain individual freedom. . . . The opposite view is based on the notion that a creative new order emerges unpredictably from spontaneous self-organization."
—R. Stacey [1996], p. 274.

simulation would create agents governed by simple rules, incorporate those agents into an environment governed by an overall fitness *mission*, and let them generate and then test a range of increasingly fit solutions. A-Life simulations create a range of solutions, many quite novel. Because of evolutionary selection pressures, the fittest solutions tend to be very robust. A-Life experimenters have created computer programs by first creating small sections of code, defining a fitness mission, and then letting the program mutate and evolve on its own.

Emergence is a central theme of complex adaptive systems theory and possibly the most important theme when looked at from a management perspective. The challenge is how to best utilize this knowledge to enhance our ability to manage software development.

Characteristics of Complex Adaptive Systems

The science of complexity, that is, the science of complex adaptive systems themselves, provides a foundation for a deeper understanding of organizational dynamics. The panoply of human-centered management practices—collaboration, self-managed teams, empowerment, distributed control, learning organizations, and participatory management, for example—lack fundamental guiding principles. While a natural correspondence exists between the scientific concept of survival of the fittest in nature and aggressive business competition, there has not, until now, been a set of scientific concepts on which to base a modern view of organizations. This new science of CAS strengthens the legitimacy of modern management practices, but even more importantly, provides new insights into improved management practices and organizational design.

Four of the most prominent scientists in the field of complex adaptive systems, all principal figures in the Santa Fe Institute (organized to study CAS), are John Holland, Professor of Computer Science and Engineering at the University of Michigan; W. Brian Arthur, Coopers and Lybrand Fellow and former Dean of Economics at Stanford; Murray Gell-Mann, Nobel prizewinner in Physics for the discovery of subatomic quarks; and Stuart Kauffman, a biologist widely known for his arrival-of-the-fittest evolutionary theories. While highly regarded scientists in their respective fields, all have also been instrumental in expanding the scope of CAS from the purely scientific biological realm

"Within science, complexity is a watchword for a new way of thinking about the collective behavior of many basic but interacting units, be they atoms, molecules, neurons, or bits within a computer. To be more precise, our definition is that complexity is the study of the behavior of macroscopic collections of such units that are endowed with the potential to evolve in time. *Their interactions lead to coherent collective phenomena, so-called emergent properties that can be described only at higher levels than those of the individual units."*
—P. Coveney and R. Highfield [1995], p. 7.

of brains, immune systems, cells, and ecologies, to the realm of human cultural and social systems, such as businesses and political parties.

Two questions must be answered if we are to understand the major issues of CAS and its application to organizations. The first is, What is a complex adaptive system? The second, What are characteristics of the environments in which complex adaptive systems live?

John Holland addresses the first of these two major issues by defining four common properties of complex adaptive systems. By Holland's definition, a CAS is an ensemble of independent agents, who exist at multiple levels of organization, who anticipate the future, and who form groups that occupy diverse niches (Holland95).

As Holland asserts, the first common, major property is that each CAS is an ensemble of independent agents. In a brain, the agents are nerve cells; in an ecological system, the agents are species; in an economic system, the agents are companies. Agents themselves have several important characteristics: Their interactions are nonlinear—the whole is more than (or at least different from) the sum of its parts. The brain, for example, is more than simply an accumulation of neurons, and memory is not merely the accumulated effect of some well-understood property of individual neurons, but results from the complex, but undivulged interaction of those neurons.

Agents in a CAS are not controlled in some centralized fashion. Control tends to be highly dispersed, yet order emerges anyway. Order is not externally imposed in some top-down fashion, but emerges from the interaction of agents, each seeking its own ends—an ever-changing blend of cooperation and competition.

Agent interactions are highly related to flows—from electrical flows in neurons to information flows in organizations. Agents respond to both the number of interconnections and the density of information in each flow.

"[Aggregation] concerns the emergence of complex large-scale behaviors from the aggregate interactions of less complex agents."
—J. Holland [1995], p. 11.

Holland identifies the second major property of CAS as the existence of many levels of organization, or aggregation. Each level provides the *building blocks* for the next level. Proteins, lipids, and nucleic acids form cells, which form tissues, which form organs, which form organisms. Similarly, people form departments, which form companies, which form economies.

In complex systems, agents self-organize to form building blocks, which in turn interact to form higher-level building blocks. Each of these building blocks may then be rearranged for a variety of different aggregates. For example, the heart is a common building block in birds, reptiles, and mammals. Nature uses the "heart" building block as a reusable component when building new species. Developers assemble new applications from reusable objects. Complex systems use building blocks—from organs to ideas—to construct ever-more-complex components, which then begin the cycle again.

The third major property of CAS is that agents *anticipate* the future. Their actions are based on a system of internal rules. An internal rule can be as simple as one for *E. coli* bacteria seeking food by moving in the direction of a glucose gradient, or as complex as a business's policy rule that anticipates future economic conditions. Holland refers to these as "internal models" that contain implicit or explicit assumptions about the way the agent's world operates. Internal models are the building blocks of behavior, and are tested, refined, and rejected as the agent investigates its environment.

The fourth, and last, in Holland's list of major properties of a complex adaptive system is the presence of many niches in an environment, each capable of being exploited by a particular agent. These niches give rise to diversity, as agents adapt in order to survive and thrive. Diversity expands as agents create niches, whether the niches occur in the biological flora and fauna of the Galapagos Islands or in the Internet market.

Orderly, Chaotic, and Complex Realms

John Holland details a second major issue in complex adaptive systems, that is, it is imperative that we understand the nature of environments in which the agents live and adapt. Complex adaptive systems comprise individual agents who in turn aggregate to form the total environment, for example, an individual company partially defines a market in which the company also competes.

This aggregate environment is called a fitness landscape—the array of all possible fitness strategies available to the agents in the environment. As each agent attempts to survive and thrive, it follows its own internal model of fitness. The aggregate of each agent's actions becomes the environmental landscape, analogous to a mountain range. The range may be smooth and undulating, like the Appalachian mountains, or rugged and precipitous, like the Rockies.

The evolution of a particular agent—say a company—is its journey across the fitness landscape, as it looks for the highest peak. However,

"If nature uses certain principles to create her infinite diversity, it is highly probable that those principles apply to human organizations."
—M. Wheatley [1992], p. 143.

"In a community of evolving systems, such as those in . . . an industry or an international economy, the suitability of a survival strategy pursued by one system—its fitness—depends upon the survival strategies being employed by other related systems—their fitness."
—R. Stacey [1996], p. 81.

in this search, it is easy to become locked into suboptimization on a subsidiary peak. Continual optimization of a certain strategy moves the firm up a peak, but finding the highest peak requires adapting techniques that search the overall landscape, often by taking seemingly inappropriate paths (a jump from one peak to another is often perilous). In emerging industries and with complex projects, even exploring which peak to ascend can be daunting. Similar to a company's search for the highest peak, a project team searches across a fitness landscape consisting of all its possible mission profiles (combinations of scope, schedule, resources, and defect levels).

The three basic categories of fitness landscapes, discussed in terms of CAS theory, are orderly, chaotic, and complex. The complex area is a transition zone, called the edge of chaos. Understanding these three zones, and in particular, understanding which one a company or project is operating in, is vital to success.

Orderly, or stable, environments are in equilibrium. They change slowly enough for linear approximations to work—cause and effect are alive and well. Stable environments are necessary; they actually provide a platform from which change can venture forth. However, stability has a tendency to breed rigidity and resistance to any changes, even those needed to survive. It may be that the most devastating problem with stable environments is that their agents *forget* how to change. Changing, like anything else, depends on doing it every now and then so the skills are not lost. In stable environments, change occurs so infrequently that agents forget how to change; therefore, it is an agonizing process when it does happen.

Chaotic environments are, for all practical purposes, random. There may be some obscure, hidden pattern, but it is of interest only to theoreticians. Chaotic environments are patternless and disorganized. In chaotic systems, slight variations can obscure any predictions of the future.

The term "the edge of chaos" was first used by Chris Langton, whose work also gave rise to the scientific field of Artificial Life. According to Langton (Waldrop92), maximum complexity is supported in this transition zone. Biological evolution in particular seems to demonstrate a proclivity for seeking the transition zone—maintaining enough control to keep from spinning off into chaos, yet enough spontaneity for creativity and enough innovation to enable it to adapt to changing environments. It is a difficult balance. In organizations, forces either push for stabilization or push for less order; they rarely stay in the middle. Understanding that the *edge* is there, understanding that it is where emergence happens, and understanding that people will be uncomfortable with it are all key to exploiting its benefits.

In human terms, being chaotic is analogous to being psychotic; being stable is analogous to being comatose. So, the trick is to lead a project team away from the familiar and the stable, toward chaos, but not all the way. Success goes to those who can *hold* anxiety, who can attune themselves to paradox and uncertainty, and who can sustain creative tension. Innovation, creativity, and emergent results are born in the transition zone at the edge of chaos.

However, theory is not enough. Practices, processes, policies, and frameworks are needed to convert the theoretical into the actionable. In software development, the most important of these frameworks is how organizations view their delivery life cycle. To survive in complex environments, both organisms and organizations must learn to adapt *at least as rapidly* as the environment itself is changing—traditional static life cycles are not adaptive enough.

"I suspect that the fate of all complex adapting systems in the biosphere—from single cells to economies—is to evolve to a natural state between order and chaos, a grand compromise between structure and surprise."
—S. Kauffman [1995], p. 15.

The Adaptive Development Model

Life cycles are used to organize the software development effort. Each type of life cycle sends a different message to the development teams: a message of relative certainty in the case of a Waterfall Life Cycle, or one of learning as the project progresses, as in a Spiral Life Cycle. Each message embodies fundamental assumptions about software management. After providing a brief overview of several classic life cycle approaches, the remainder of this chapter focuses on the Adaptive Development Model Life Cycle, which has evolved from historical spiral and iterative roots and which embodies the underlying principles of complex adaptive systems.

The fundamental difference between the Adaptive Development Model Life Cycle and other iterative life cycles (spiral and evolutionary) is that the latter still operate on a belief in imposed order and cause-and-effect engineering. They attack high change in the environment by reducing cycle times and improving feedback practices, but they do not provide the systems engineer with the mental switch to a belief in emergent order. The Adaptive Development Model Life Cycle is also explicitly a component-based rather than task-based approach.

The Evolution of Software Life Cycles

Just as physicists and economists have begun to deal with the unpredictability of their sciences, software developers are revising their development practices to address the growing uncertainty of their projects. Described below are three broad stages in this progression of systems development life cycles—waterfall, evolutionary, and adaptive.

Waterfall Life Cycle. For many years, the dominant software development life cycle was the waterfall approach, which is characterized by linearity and predictability, with a modicum of feedback thrown in for good measure. The waterfall approach, illustrated in Fig. 2.1, produced the bulk of legacy systems still in use today. While waterfall methods brought some stability and order to development efforts, their shortcomings are becoming increasingly apparent and many software development organizations are lost in the swamp created by too much water over the fall.

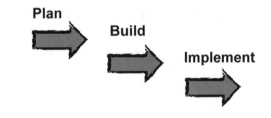

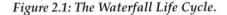

Figure 2.1: The Waterfall Life Cycle.

Evolutionary, Spiral, and RAD Life Cycles. Based on the pioneering work of Tom Gilb and Barry Boehm (Gilb88, Boehm88, Boehm98), evolutionary and spiral life cycles began to emerge in the mid-1980's. Gilb used the term "evolutionary" to describe an iterative approach with small, well-planned development cycles. Boehm's spiral model explicitly incorporated the concept of driving development plans based on risk.

Evolutionary, spiral, and RAD life cycles all are iterative, as shown in Fig. 2.2. They address uncertainty by taking smaller steps and testing the results. Each step's deliverables are included in an overall project plan that is then usually revised at each subsequent iteration. Life cycle variations are sometimes given different names based on how much architectural and design work is done initially and whether or not each iteration involves actual implementation.

While these models have moved into the mainstream, many practitioners have not changed their deterministic mindset. Long-term predictability has been abandoned for short-term predictability, but it is predictability nonetheless. For example, Gilb's recent work entails detailed component planning and great precision in specifying requirements.

Another variation in iterative life cycles is Rapid Application Development, also referred to as RAD/Prototyping. RAD life cycles often contain aspects of evolutionary and spiral approaches, with some combination of time-boxing, a high degree of user involvement including Joint Application Development (JAD) sessions, and powerful development tools. While many RAD/Prototyping approaches are more adaptive in practice, they still lack articulated adaptive concepts. Additionally, RAD approaches are not usually recommended for larger projects.

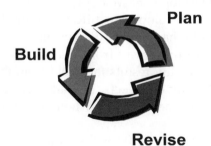

Figure 2.2: The Iterative Life Cycle.

The Adaptive Development Model Life Cycle. The life cycle for the Adaptive Development Model, a key component of the overall Adaptive Software Development approach, is built on a different world view, one of adaptation rather than of optimization. While the approach is iterative as in the evolutionary model, the phase names reflect the unpredictable realm of increasingly complex systems. Adaptive Software Development goes further than its evolutionary predecessors in three key ways: First, ASD acknowledges the reality of uncertainty and change and therefore does not try to manage projects through precise prediction and rigid control strategies. Rather than control, ASD's strategy is more subtle—to bound, direct, nudge, or confine, but not to control.

The second unique feature of ASD is that it explicitly encourages a culture of *emergent order* rather than *imposed order*. A strategy that depends on emergent order goes beyond altering life cycles or phase names, although sometimes the difference can be subtle. For example, as the environment changes, those using a deterministic model would look for a new set of cause-and-effect rules, while those using the adaptive model know there are no such rules to find.

The third characteristic of ASD is that it is explicitly component-based rather than task-based. From the perspective of management strategy, ASD focuses on results (components) and the defined constraints (quality characteristics) of those results. The development team then executes processes, or tasks, to produce the components within the constraints. While being component-based is not unique to ASD, it is an important differentiator in the era of highly process-driven strategies.

The Adaptive Development Model Life Cycle (also called, simply, either the Adaptive Development Life Cycle or the Adaptive Life Cycle) is iterative, but it has another dimension. The secondary arrows in Fig. 2.3, shown leading away from the iterative circle, represent breakout ideas that identify results far afield from the original project mission profile. Adaptive projects, like adaptive organizations, are open to possibility—one never knows where the next innovative product will take seed.

Overall benefits resulting from the Adaptive Development Life Cycle include the following:

- Applications evolve in response to periodic feedback, resulting in a close match to customer requirements.

- Changing business needs are accommodated more easily.

- The development process adapts to the specified quality profile of the product.

- Customer benefits are generated earlier, for example, because the customer gets the application more quickly and can use it to increase revenue.

- The risk that major failures will occur is reduced.

- Customers gain early confidence in the project.

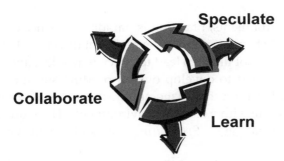

Figure 2.3: The Adaptive Development Life Cycle.

Speculate–Collaborate–Learn

Examine the activities named in the three phases of the Speculate–Collaborate–Learn cycle, even briefly, and you will notice obvious overlap. For example, it is difficult to collaborate without learning, or to learn without collaborating. The phases are purposely nonlinear and overlapping to emphasize ASD's approach to managing uncertainty.

For many project leaders and project teams, Adaptive Software Development is a terrifying prospect. First, the traditional supporting pillar of cause and effect is knocked away, so teams are never sure what needs to be done next. Then, the team is directed to meet deliverable goals that are admittedly unclear. When team members feel anxiety about the seemingly inefficient quest for solutions that is typical of ASD, they may be told that high anxiety is part of working with uncertainty. And finally, when a successful product emerges, it seems almost accidental. This approach is not for the timid, but there are organizations actively practicing some form of Adaptive Software Development—Microsoft is one such example. Despite tell-all books that describe Microsoft's techniques and practices—see, for example, Michael Cusumano and Richard Selby's *Microsoft Secrets* (Cusumano95)—Microsoft prospers because at the core of these very techniques and practices lies a management philosophy that espouses abandoning determinism and embracing an adaptive environment's complex nature.

Speculate

"The ability to plan for what has not yet happened, for a future that has only been imagined, is one of the hallmarks of leadership of a Great Group. . . ."
—W. Bennis and P. Biederman [1997], p. 40.

In complex environments, planning is paradoxical, since outcomes are unpredictable. Yet endless, aimless experimentation on what a product should look like is not likely to lead to concrete results either. *Planning*, whether it is applied to develop overall product specifications or to articulate detailed project management tasks, is too deterministic an activity to associate with a complex environment. The word carries too much historical baggage. *Speculating* seems closer to what actually occurs.

When we "speculate," we define a mission to the best of our ability, but our choice of the word indicates that in some important dimensions of our mission statements, we are more than likely wrong. Whether it is technology changing or competitors coming out with a better mousetrap or our own misreading of our customers' needs, the probability of mistakes is high. So, let's be honest, postulate a general idea of where we are going, and put mechanisms in place to adapt—to *explore* the territory.

In a complex environment, following a plan produces the product you intended, just not the product you need.

Viewing a project as an exploration, as a learning journey, is actually contradictory to most current practice. Figure 2.4 shows two deviant project paths. From a traditional perspective, projects should go from the current state (point A) to the planned result (point B) as indicated by the dashed line. But most traditional managers acknowledge that there will be deviations in the execution of the plan as shown by the zigzagging path between the two points. The assumption underlying this traditional approach is that

Deviations from plan are mistakes that must be corrected.

Traditionalists view deviations as errors to be eliminated. Even worse, they view those who make the mistakes as lazy, careless, inattentive, or maybe even dumb. Deviations in project status reports cause such concern, in part, because the entire psychology of most project managers is one of adhering to the plan with the fewest deviations possible.

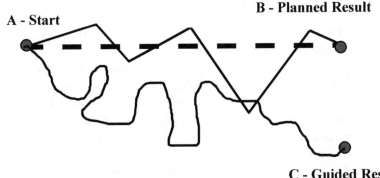

Figure 2.4: Planning versus Guiding.

The squiggly, second path in Fig. 2.4 represents a different model of development, which reflects a correspondingly different mental model. Point A (the current state) and point B (the initial goal) are the same, but the team *expects* to end up at point C, even though team members don't know where C is when they set out from A. Their outlook is that

Deviations guide us toward the correct solution.

The second model is not an anti-planning model, but a realistic one. It tells us that we, as messengers, can't know enough in the beginning, no matter how smart we are, especially as the market becomes less predictable. To compensate for the lack of certain knowledge, team members need to establish both new development approaches and better learning practices.

Planning means defining a specific goal and the path (the tasks) by which to achieve it—planning restricts exploration, implicitly if not explicitly. Speculating means defining a broader goal and allowing for alternative paths—speculating explicitly recognizes uncertainty and leaves open more possibilities.

Adaptive agents explore fitness landscapes for better solutions. This adaptive exploration begins with speculating, which involves both focusing and de-focusing activities. Two primary *focusing* practices are defining (speculating on) a mission and developing a detailed Adaptive Life Cycle plan.

Speculating on a Mission

The overriding criterion for project success is delivering results—it is not the software engineering techniques used, or the process improvement method, or the philosophy. In determining if software product organizations are successful, regardless of whether they are software vendors or IT organizations, our first criterion should be whether products are delivered. It is easy to get caught up in designing architectures, putting infrastructures in place, building teams, and improving processes, and then to lose track of the real goal.

I use the word *mission* in this book to include a wide variety of focusing statements, such as vision, objectives, goals, and even high-level requirements and constraints—anything that helps the project team define the desired results of the effort at a summary level. A mission's purpose is to direct work activities toward an end result.

Defining the mission is all about the "what." What is the product being produced? What are the desired quality attributes of the product? What is quality itself, as applicable to the product? The focus is on the results—not just on product features but on all aspects of value: scope, schedule, defect levels, resources. As my colleague Lynne Nix reminds me, the value of the mission is not only the focus it provides to the project team, but also the focus it gives to managers who have a tendency to sidetrack projects. The Adaptive Life Cycle keeps the project focused by providing iterative versions of the product—artifacts of the actual application—as the primary deliverables.

Defining the mission is not only about defining the "what." It is also about sharing the mission. The idea of shared values has been written about in management literature for some time, but it is often dismissed in practice as being too soft and mushy, as being too intangible to be useful. Despite the number of books, conference presentations, and articles written on the topic of having a good *mission* statement, the message of sharing is getting lost. Developing a good set of mission statements isn't easy, but the techniques are reasonably well-known. Developing a shared sense of the mission among team members—including customers, developers, management, and vendors—is a much more difficult task.

Collaborate

Managing in a complex environment can be scary, but it can also be stimulating. If we can't predict, or plan, then we can't control (in the traditional management sense). If we can't control, then a significant set of current management practices are no longer operable or, more specifically, are only operable for those more *predictable* parts of the development process (for example, configuration management).

Collaboration creates organizational *emergence*—that *aha!* property separating the great from the merely good. At times, emergent properties seem almost mystical, but in field after field, from physics to cellular automata to software development projects, emergence has been demonstrated. We have all experienced emergence on some special project, but it somehow seemed accidental, or at least unruly and undependable—not something to count on in a crunch. The study of complex adaptive systems is proving wrong the view that emergence is some *spooky* property. It may be somewhat unpredictable, but emergence is far from accidental.

Journalist and writer Michael Schrage defines collaboration as "an act of shared creation and/or shared discovery" (Schrage89, p. 4). In contrast to collaboration, communication is passive—it transfers information, with an intent to inform. Collaboration is active—it requires active participation, with an intent to add value.

Creative collaboration thrives on diversity, rich relationships, unfettered information flow, and good leadership. At the feature-team level, creative collaboration is manifested as interpersonal dynamics and collegiality. Culture and structure drive it at the organizational level.

"[S]uccessful collaboration is the science of the possible."
—W. Bennis and P. Biederman [1997], p. 131.

Learn

The popularity of books like Peter Senge's *The Fifth Discipline* have given credence to the concept of a learning organization, an organization capable of changing with the times (Senge90). Learning organizations have the ability to critically examine themselves, in particular to examine the assumptions or "mental models" that underlie their processes and practices.

Collaborative activities build products. Learning activities expose those products to a variety of stakeholders to obtain feedback on their

"Real learning gets to the heart of what it means to be human. . . . Through learning we become able to do something we never were able to do. Through learning we reperceive the world and our relationship to it."
—P. Senge [1990], p. 14.

value. Customer focus groups, technical reviews, beta testing, and postmortems are all learning practices exposing results to scrutiny.

Learning means gaining mastery through experience. In an adaptive environment, learning challenges all stakeholders—developers and their customers—to examine their assumptions and to use the results of each development cycle to adapt the next. In the adaptive approach, cycles are short, so that participants learn from small, rather than large, mistakes.

"The software industry has a capacity for almost infinite self-delusion."
—C. Jones [1996], p. xxvii.

The Adaptive Life Cycle's structure *en*courages, rather than *dis*courages, learning. It is quite possible to execute a serial Waterfall Life Cycle project without gathering or listening to feedback from either customers or developers (the reason for this is that the waterfall mental model exudes confidence and knowledge—a sense that "we know what to do and how to do it; let's get going"); however, the adaptive model is more humble. It forces us to admit up front our lack of complete knowledge, and it compels us to build in more comprehensive feedback mechanisms.

> **The overriding, powerful, indivisible, predominant benefit of the Adaptive Development Life Cycle is that it forces us to confront the mental models that are at the root of our self-delusion. It forces us to more realistically estimate our ability.**

Facing our mental models is an emotional experience, a fact that seems to get lost in academic discussions about learning organizations. People don't cling to mental models or assumptions or beliefs because they are obstinate. They embrace them because, deep down, they are passionate about them. Understanding the emotional basis for mental models is an important part of becoming an adaptive, learning organization.

Working in a Complex Environment

Sometimes, at least to outside observers, the activities of an adaptive group and a chaotic group may look very similar, but nothing could be further from the truth. There are three main differences between chaotic groups and adaptive ones. First is the difference in their mindsets and how they operate. An adaptive group knows the difference between chaos and complexity. A chaotic group thinks the only alternative to chaos is boredom.

Second, adaptive group members have a shared mission that keeps them on track. Group members may not know their specific destination, but they have a good sense of direction, which keeps being refined at each iterative cycle. By the last cycle, they understand the specific deliverables and the contents of those deliverables. Chaotic groups never converge on results; they randomly fly off in unanticipated directions.

Third, an adaptive group understands when rigor is needed, whereas the chaotic group decries rigor as bureaucratic. A chaotic group is on a downward death spiral. An adaptive group converges on results. For example, an adaptive group knows that the first remedy for a truly chaotic project is to instill some stabilizing practices.

Adaptive groups understand the need to balance optimizing (linear, engineered, precise) practices and adaptive practices. They know that these two views of software management have complementary, not adversarial, positions. In fact, there is not a clean demarcation between the two worlds, but a spectrum across which both views must shift.

With the increased pace of competitive change, the reduction of product delivery cycles, and the explosion of new technology, more industries must compete in an environment of increasing returns. For software producers, the need for adaptive development techniques arises when there are many independent agents—developers, customers, vendors, competitors, stockholders—interacting with each other too fast for linear cause-and-effect rules to be sufficient for success.

As businesses shift from manufacture of resource-based products to that of knowledge-based products, successful software development will help determine which companies prosper in an increasing-returns environment. Development will need to migrate toward an adaptive viewpoint to meet the demands of an unstable, complex, and chaotic world.

From the perspective of complex adaptive systems, there are three concepts fundamental to working in an uncertain environment. First is the concept of moving from a deterministic, linear, cause-and-effect view of the world to one based on adaptation, nonlinearity, and increasing returns. Committing to an adaptive mindset is a commitment to a learning organization in which learning means not just acquiring knowledge, but also understanding how organisms and organizations use mistakes, short feedback cycles, and critical self-examination.

Second is the concept of emergence, which is key to this new, adaptive mindset. If *arrival of the fittest* supplants *survival of the fittest,* then practices and tools to build dense networks of highly connected, collaborative agents and teams must receive as much management attention as implementing the next Java development kit. Self-organization is not a controlled, top-down phenomenon but one arising from the messy interaction of the individual agents seeking a better way.

Third is the importance of information flow, the raw material of self-organization and emergence. Emergent results, those greater than deterministic practices can explain, are generated from the intense interaction of diverse agents. In software development, those interactions are flows of information. Balancing the content and number of information flows is an important activity in creating an environment conducive to emergence.

Complex adaptive systems are defined by agents and their environments—climbers and mountains, tree frogs and rain forests, software engineers and markets. If they have some underlying symmetry, if they respond similarly as complex living systems by adapting through a balance of cooperation and competition, then both our Monumental and Accidental management edifices are at risk. It may seem that an adaptive mental model is more compatible with an Accidental development approach than with a Monumental one. In fact, it actually changes the perspectives of both Monumentals and Accidentals.

The dangers of a Monumental mindset lie in rigidity and in its failure to adopt changes needed to prosper. Organizations in this mode often point to implemented changes as a defense against the charge of rigidity, but these changes are more often optimizing rather than adaptive. From an adaptive perspective, the Monumental mindset can learn

- to create an environment in which exploration of the complex transition zone at the edge of chaos is possible

- to purposely destabilize in order to explore new territory

- to use techniques that encourage adapting rather than optimizing and to become more comfortable with them

- to foster practices and principles that move toward a participatory, relationship-building environment and away from a specialized, interchangeable-parts view

The dangers of the Accidental mindset can be more subtle, especially to those who apply this classification to themselves. The danger in these organizations is that they will fall into chaos, into complete disorder. Their defense mechanism is hoisted as the twin flags of creativity and individualism. From an adaptive perspective, the Accidental organization can learn

- to create parts of the environment in which exploration of the transition zone moves from chaotic to more rigorous

- to purposely stabilize in order to reap the benefits of innovation

- to recognize the difference between anarchy and adaptation and to understand that unfettered information flow leads to chaos

- to understand that viewing people as independent individualists is just as detrimental to adaptation as the interchangeable-parts view

Whereas Monumental organizations restrict information flow, Accidental ones revel in its boundless expansion. In many ways, but for different basic reasons, neither Accidental nor Monumental organizations are very good at creating environments conducive to emergence.

Summary

➢ The Adaptive Conceptual Model focuses on agents, environments, and emergence.

➢ A gentle hike to a beautiful alpine lake, a moderately difficult technical rock climb, a world-class sport-climbing route—each represents a particular class of environments. Software projects reflect similar differences in level of difficulty.

➢ Max the athlete, Bruno the hulk, Julia the former gymnast—each represents a class of climber, or agent, with distinct skills, emotions, and attitudes. Software developers and development teams reflect similar differences.

➢ Climbing is about the interaction between the environment and the climber. Software development is about the interaction of the development team and its environment—the market. For the most part, the mountain stays still for the climber; software development teams are not always so lucky with their environments.

➢ Emergence is a property of complex systems that generates extraordinary results from the interaction of pieces. It is not the result of deterministic planning, but of creating an environment in which results emerge from a diversified mix of ingredients. Those results, however, may not be the ones anticipated.

➤ Complex Behavior = Simple Rules + Rich Connections

➤ Complex adaptive systems consist of independent agents, who can be aggregated at many levels, who anticipate the future, and who fill diverse ecological niches.

➤ There are three categories of fitness landscapes, the environments in which complex adaptive agents live—stable, chaotic, and complex. Creativity, innovation, and emergence thrive at the edge of chaos.

➤ Software development life cycles have evolved from the serial, Waterfall Life Cycle model to the iterative, evolutionary (or spiral) model, to the iterative, emergent, Adaptive Life Cycle—Speculate–Collaborate–Learn.

➤ Speculating is analogous to fuzzy planning. There is a mission, even one specifying outline requirements and plans for each development cycle, but rather than being viewed as a definitive target, the plan is viewed as a direction to be explored.

➤ In a deterministically planned project, deviations are viewed as mistakes to be corrected. In an adaptive project,

Deviations guide us toward the right solution.

➤ The overriding, powerful, indivisible, predominant benefit of the Adaptive Life Cycle is that it forces us to confront the mental models we harbor as the root of self-delusion. It forces us to more realistically estimate our understanding.

➤ Collaboration means working together to produce a shared result. It is an act of adding value to a product. Communication means the passive transfer of information.

➤ Learning means modifying our mental models of the world as a result of experience and new information. It is about the ability to admit to, and thereby profit from, our mistakes as well as our successes.

➤ Different environments require different solutions. Stable environments are the province of optimizing approaches. Complex environments require adaptation.

Part 2

Part 2

CHAPTER 3
The Project
Mission

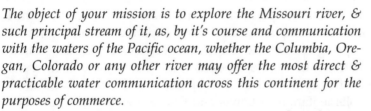

The object of your mission is to explore the Missouri river, &
such principal stream of it, as, by it's course and communication
with the waters of the Pacific ocean, whether the Columbia, Ore-
gan, Colorado or any other river may offer the most direct &
practicable water communication across this continent for the
purposes of commerce.
   —T. Jefferson, as quoted in S. Ambrose [1996], p. 94.

With these words, Thomas Jefferson launched the Lewis and
Clark expedition of discovery, the most notable event in
American exploration. For thirty months, these words drove
the members of the expedition on—through brutal physical challenges,
frightening Indian encounters, isolated winters, and life-and-death
decisions. Writing a mission statement is one thing; understanding the
scope, meaning, subtleties, ambiguities, and limits of a mission is
another thing altogether.

Meriwether Lewis was Thomas Jefferson's secretary for several
years before the expedition began. The men knew each other well,
talked long about the expedition and its goals, and exchanged extensive correspondence on various aspects of the journey. At a time when

the Federalists were attacking Jefferson for his purchase of the Louisiana territory, Jefferson and Lewis shared a vision of a unified continental United States. They agreed on the purpose of the expedition, viewing it as one means to achieve their grander continental vision. Lewis passed their vision on to William Clark and the other explorers—a vision consisting not only of words, but also of the passion and inspiration to bring the words to reality.

A broad mission statement is important, for it gives purpose and meaning to a task, but it needs to be supplemented with specifics. Jefferson's mission statement gave specific instructions about mapmaking: ". . . you will take careful observations of latitude & longitude, at all remarkable points on the river, & especially at the mouths of rivers, at rapids, at islands . . ."; about Indian ethnology: "to learn the names of the nations, and their numbers, the extent of their possessions, their relations with other tribes . . ."; and about flora and fauna: "to notice and comment on the soil, the plant and animal life, . . . dinosaur bones, and volcanoes" (Jefferson, in Ambrose96, pp. 94, 95).

The thoroughness of the mission statement was critical to the success of the expedition. Because the explorers would be out of contact with the rest of the world during most of the trip, the mission statement needed to be specific and yet flexible enough to allow the leaders to make key decisions as their party explored the unknown. (Indeed, by the spring of 1806 and almost two years into the journey, many people had given up on the expedition as "lost in the wilderness.") Jefferson's instructions focused on objectives and broad directives but left the implementation to Lewis and Clark. Lewis and Clark didn't succeed because of a good mission statement, but they would not have succeeded without one.

Good mission statements don't come easily. Jefferson's mission statement took years of study and represented the collected wisdom of many colleagues. How many project teams, when asked if they have a mission, would answer, "No. We just fumble around all the time"? Most team members probably think they have some sort of mission, but a good mission is not a thing, not words on a page or on a flip chart. A good mission is shared passion. At its most basic, a mission is a product goal worth striving for. At its best, a mission touches each team member's sense of a wider purpose beyond producing some *thing*.

There are three steps to fashioning a statement of mission. The first is to *identify the mission*, to understand what constitutes a good mission. The second is to *create mission artifacts*, to define specific mission documents and to develop their contents. The third step is to *share mission values*, to go beyond the words on paper or in digital images so that each team member shares the passion and purpose of the mission.

Identify the Mission

A good mission statement is many things, particularly in a complex environment. It must be concrete, yet still invite people to speculate on different scenarios. It must also facilitate the flow of information. The objective of a development effort is to build a usable software product, so a good mission statement needs to be specific. However, if it is too specific, the project team may be overly constrained. A mission statement needs to help the team converge on a solution while still keeping the team open to divergent innovation. A mission statement needs to direct or define the scope or boundaries for the effort, not detail the final outcome.

Characteristics of a good mission statement are threefold. The first is that it will establish a sense of direction. At a project or product level, a mission establishes a framework for action, a scope for what is to be accomplished, and a theme for design.

A second characteristic is that a good mission statement is inspirational. People's best efforts arise from their being passionate about what they are doing. A mission statement answers questions such as, Why do I care about this project? Why should the team care about this project?

There is a third aspect to a mission statement. It guides implementation, suggesting how the project should be approached and providing a framework for decision-making. Steering development processes and practices requires knowing what one is steering toward—the mission. Engineering is in some fundamental way a process of synergy and trade-off, of trying to derive a design incorporating often disparate requirements and making technical trade-offs based on business criteria.

Hundreds of decisions—some major, many minor—are made during a software project. Particularly in an adaptive, iterative project where team members and clients are encouraged to learn and change, a lack of boundaries defined by a mission statement would create chaos. Decisions must be made. Trade-offs are a necessity for moving forward. High rates of change mean that even more decisions must be made, and quickly. Good mission statements guide decision-making.

"[U]nity is the master principle of great art. And I have seen over and over that unity is the master principle of great software. . . . The theme of your software is the dominant idea that constitutes the basis of the design. . . . You've got to have a purpose for your product, and 'unity of purpose' is a good phrase to describe the impact of having a theme."
—J. McCarthy [1995], pp. 80, 81.

"A mission is a sense of purpose that lures you into your future. It unifies your beliefs, values, actions, and your sense of who you are. . . . Most of all, a mission is fun."
—J. Andreas and C. Faulkner [1994], p. 80.

It is not enough to have a two-sentence vision statement: Different *levels* of mission are needed to make different levels of decisions. Broadsweeping, generalized mission statements are useless when detail-level decisions must be made. Narrow, precise mission statements don't allow for innovation and flexibility. Part of the ability to develop a good mission statement lies in the development team's ability to understand this ambiguity.

In a complex environment, a mission statement is a speculation about the future, and everyone involved must remember that this speculation is the best guess of the future available at the time, something to be continually tested against reality and adjusted accordingly. At a conceptual level,

A mission statement facilitates collection of information that is relevant to the project's desired result.

Any product development effort is, at its core, the gathering, analyzing, and reconfiguring of information. One of the defining characteristics of a complex environment is the high rate of information flow. In a stable environment, information flows slowly and predictably. In a complex environment, information bombards the project team from numerous and often unforeseen sources. How do team members extract the information they need from the vast possibilities?

Industrial robots use "rules" to manufacture automobiles. Insurance service administrators use well-defined "processes" to initiate a policy. Chess players use "patterns" of play to outsmart their competitors. In software engineering (and in business), the word "process" has become synonymous with a carefully controlled procedure for arriving at planned results. What we need is a new word to connote insight without certainty—and the word becoming more widely used is "pattern." A design pattern is a framework for helping someone transfer knowledge, but it is not a recipe to be followed by rote. Success on complex projects requires that we comprehend the difference between processes and patterns. The first prescribes activities, the second organizes thinking. A mission statement should suggest a direction rather than a destination and must help us select the patterns that will increase our chances of success.

To me, the word "process" suggests a mechanical world, while the word "pattern" suggests an organic one. Adaptive Software Development is not a series of processes, but an assemblage of patterns that can assist development teams in their thinking but that also have limitations.

Innovation and creativity are not the result of processes, but of patterns. Management and organizational patterns are emergent; we cannot always explain exactly how they work, but we recognize that they often (although not every time) produce results.

A mission statement helps us navigate through and understand a complex project environment and facilitates the collection of information. Because of the high information-flow rate in a complex environment, we cannot anticipate exactly what a team will need. However, by identifying a credible direction and instituting appropriate patterns of inquiry, we should be able to collect relevant information to produce the correct product. If the direction is too broad, we will find that processing all the captured bits will be too time-consuming. If the direction is too narrow, there will not be enough diversity to build a viable product. In short, to identify the mission, software project management must focus on direction and it must establish appropriate priorities to facilitate reaching that goal.

A Need to Focus

In the July 1995 issue of *American Programmer* magazine, several well-known authors supported the opinion that, to be successful, a project must meet *all* of the following criteria. The project must

- meet business objectives

- meet quality expectations and requirements

- stay within budget

- meet its time deadline

- deliver actual benefits

- provide the team with professional satisfaction and the opportunity to learn

With all due respect to these highly regarded authors, I view this stance on multiple criteria for measuring project success to be pure rubbish! There is no way to satisfy *all* of the above criteria on any reasonably sized project, much less on an *extreme* one. By making everything top priority, these authors leave managers no effective way to manage change. Change requires trade-offs, trade-offs require an understanding of priorities. My objection does not mean that any one of the criteria listed above is unimportant, but that we have to *choose one criterion as the most important.* By setting up such impossible multiple criteria for measuring success, we do not give truly successful projects their due.

The siren cry in marketing is Segment, Segment, Segment. In business strategy, it is Focus, Focus, Focus. Why should software management be any different? It is not, as I show in Table 3.1, which contains a matrix to help project leaders manage change by establishing appropriate priorities for desired project results.

A Need to De-Focus

"You do not navigate a company to a predefined destination. You take steps, one at a time, into an unknowable future. . . . In the final analysis, it is the walking that beats the path. It is not the path that makes the walk."
—A. De Geus [1997], p. 155.

The Speculate–Collaborate–Learn Adaptive Development Model Life Cycle, first depicted in Fig. 2.3, has small offshoot arrows that represent breakthrough or emergent learning, suggesting directions or product feature sets not planned but that could potentially improve the product. If project management has not even considered a potential project direction, it will find it difficult to accept data relevant to that direction. So concerned with reducing "scope creep," the seemingly constant change in product requirements, we as managers may fail to examine innovative new directions.

Software development in complex environments is difficult because it is not deterministic—that is, it is not easily controlled by simplistic (which must not be confused with *simple)* rules. While I encourage project team focus through development of mission statements, a singular focus will doom projects in a complex environment. Every project team goes through periods of focus—of converging on a solution—and then de-focus—diverging away from that solution because of new ideas, problems, or conflicting opinions. Both activities are healthy if not carried to extremes. It would be simple to say, "Always focus." It is much more difficult to say, "Sometimes focus, sometimes de-focus." Knowing when to focus and when to de-focus requires judgment—a skill often in short supply. Thinking of a mission as a boundary rather than as a destination, as a pattern rather than as a single point, can assist a project in this balancing act.

In the preceding chapter, a fitness landscape was defined as a three-dimensional representation of a project's success criterion. Thought of in terms of the fitness landscape, focusing is analogous to climbing a peak and striving to reach the top. De-focusing is analogous to stopping to consider the possibilities of jumping to an adjacent peak, one that offers greater potential because it is a higher peak but which also offers risk because it necessitates abandoning the current peak.

If we need to both focus and de-focus, we need to know when to do each. Determinists want rules such as, "When *x* occurs, focus. When *y* occurs, de-focus." This merely replaces one simplistic rule (Always focus) with a short set of simplistic rules. Unfortunately, no one rule, or even one *set* of rules, can be applied to all situations.

Holding anxiety, dealing with paradox and conflicting constraints, continually being poised at the edge, alternately applying techniques to converge and diverge—all are critical to maintaining an environment conducive to emergent results. Knowing when to focus or de-focus a project team is essential to successful management in complex environments; adaptive techniques can assist managers in making these judgment decisions.

One popular technique used for de-focusing is called *scenario planning*. Scenario planning requires project members first to change some basic assumptions about a product, a technology, or a market, and then to explore what this new direction would mean. Feasibility studies often contain an analysis of alternatives, but these are usually alternatives for implementation. Scenarios provide alternatives for significant product feature sets, or even for entire missions.

Monitoring the environment using alternative scenarios keeps the team from focusing too tightly. As in an ant colony that has scouts out exploring the territory, software developers must scour the terrain for the first indications of competitive danger or opportunity.

> *"Scenarios bring new views and ideas about the landscape into the heads of managers, and they help managers learn to recognize new 'unthinkable' aspects of the landscape even after the scenario exercise is over."*
> —A. De Geus [1997], p. 46.

Create Mission Artifacts

The second step in fashioning a mission statement is the creation of mission artifacts. Why this step is important is not always clear in the heat of a project, but it stems from the fact that software project failure is endemic. At the start of a project, everyone is in a hurry, either because management has already delayed six months in initiating the project and needs to make up lost time, or because some competitor has just struck, or because . . . just fill in the blank: "Full speed ahead." "Damn the torpedoes." "Don't be a wimp." The beginnings of many projects are injected with a tremendous oversupply of testosterone. Three or four months into the project, someone finally realizes that no one knows what the project is all about, and a reevaluation period ensues. The following statistics indicate how wasteful this is.

> *Our research shows a staggering 31.1 percent of projects will get canceled before they are ever completed. Further results indicate 52.7 percent of projects will overrun their initial cost estimates by 189 percent. . . . Over one-third of these same challenged or impaired projects experienced time overruns of 100 to 200 percent. One of the major causes of both cost and time overruns is restarts. For every 100 projects that start, there are 94 restarts.*
> —J. Johnson [1995], pp. 4, 5.

That 94 percent of all projects will be *restarts* is incredible! According to Johnson, we lose $78 billion a year on canceled IT application development projects. A good set of mission artifacts, developed in a feasibility study process, can reduce the losses substantially.

Mission artifacts comprise the specific documents developed for a project. While the details must be customized for each project, these artifacts need to answer three questions:

1. What is this project about?

2. Why should we do this project?

3. How should we do this project?

A concept is not a project, but a beginning, a thing to be examined and *nurtured into* a full-blown project. Many projects get in trouble because they confuse *starting* fast with *ending* fast. In most situations, taking extra time to get the project started properly pays big benefits in time saved at the end.

What is this project about? To provide context for this question, we might paraphrase Lewis Carroll's *Alice in Wonderland:* "If we don't know where we want to go, then any path will do." One primary reason for project restarts, or outright failure, is the lack of a project mission. Team members, customers, and other stakeholders need a good understanding of the project's fundamental components—its goals, objectives, scope, problem statement, constraints, and vision. A good test of whether project participants understand a project is to walk around and ask them, "What is this project about?" The more complicated the answer, the more trouble the project is in. A crisp, business-

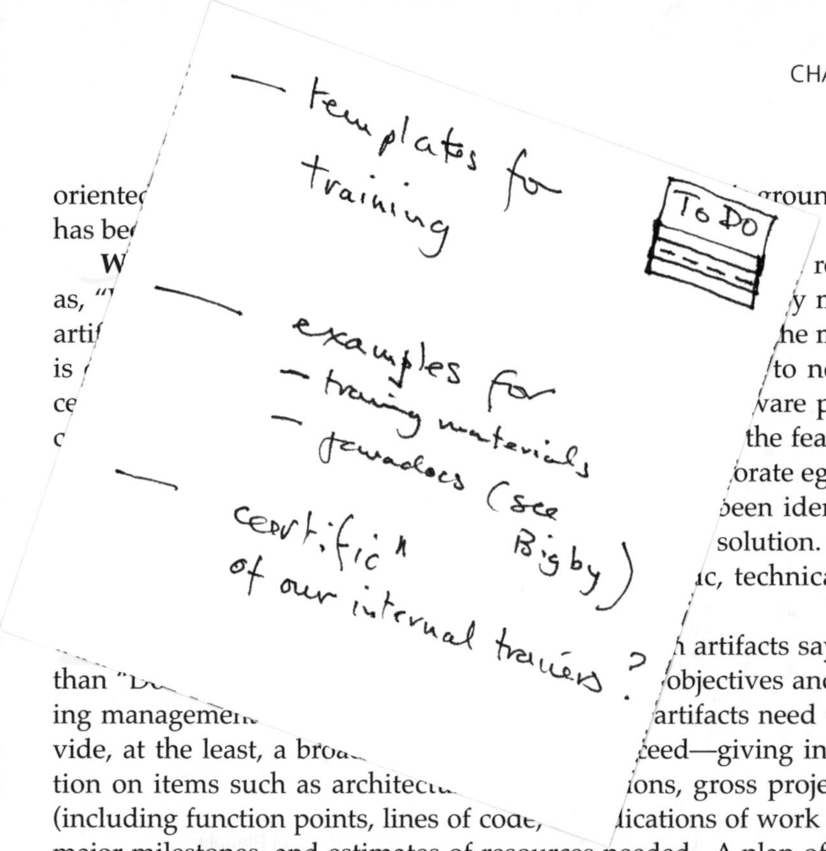

(handwritten note overlay)
— templates for training

examples for
— training materials
— javadocs (see Bigby)

certification of our internal trainers?

To Do

oriented _____ _____ ____ _____ groundwork has be_
W_ _____ restated as, "_ _____ y mission artif_ _____ he mission is _ _____ to not pro- ce _____ ware projects c_ _____ the feasibility _____ orate egos and _____ been identified, _____ solution. Feasi- _____ ic, technical, and

_____ artifacts say more than "_____ objectives and help- ing managemen_ _____ artifacts need to pro- vide, at the least, a broa_ _____ eed—giving informa- tion on items such as architect___ ___ions, gross project size (including function points, lines of code, ___lications of work effort), major milestones, and estimates of resources needed. A plan of action serves two purposes: It gives the follow-on team a direction and it forces the feasibility team into thinking about critical implementation issues early in the project.

Words such as mission, vision, goals, objectives, requirements, and theme apply to many of the concepts discussed in this chapter. I have cited different authors on what defines a mission, but I will not attempt a single-sentence definition because I believe a single sentence, or even a single document, isn't broad enough to encompass the concept of mission. However, by identifying the basic elements, the artifacts, of a mission, I show that a project's mission then becomes an amalgamation of these artifacts. My intent is not to prescribe exact mission artifacts for each and every project or organization, but to describe *types* of arti- facts from which each project's needs can be determined.

The most important mission artifacts are the project vision (char- ter), the project data sheet (PDS), and the product specification outline (PSO). They are defined briefly below, are illustrated in Fig. 3.1, and then they are discussed more fully in named subsections.

The Project Vision (Charter): The charter provides a short, two-to- ten-page definition of key business objectives, product specifications, and market positioning.

"Visions are aesthetic and moral—as well as strategically sound. Visions come from within—as well as from outside. They are personal— and group-centered. Developing a vision and values is a messy, artistic process. Living it convincingly is a passionate one, beyond any doubt."
—T. Peters [1987], p. 486.

The Project Data Sheet: As originated by my colleague Lynne Nix, this staple of her seminars is a single-page summary of key business benefits, product specifications, and project management information. The sheet serves as a document to help focus the project team, management, and customers.

The Product Specification Outline: The outline inventories the features, functions, objects, data, performance, operations, and other relevant specifications of the product at a high level.

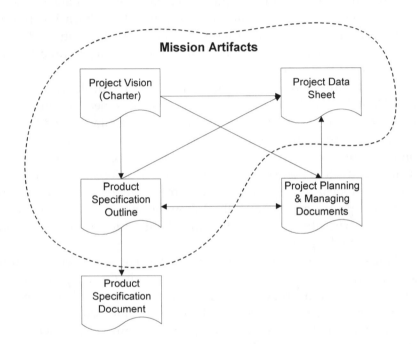

Figure 3.1: Mission Artifacts.

The Project Vision (Charter)

The project vision is recorded in a document, or artifact, which establishes a focus for the project and identifies the foundation on which to build the team's commitment. It establishes which direction to take into the fog of the unknown. It provides boundaries for the exploration phase of the Speculate–Collaborate–Learn Adaptive Development Model Life Cycle. Depending on the situation, a project vision could be stated as a single sentence or a multi-page document such as a project charter or a project feasibility study report. The specific categories of information within the vision document will be different for each organization, for each size project, and for each release of a product.

Most projects have existing historical information developed by clients, marketers, or developers. This information may range from a detailed market research study in a product company, to a project proposal and cost/benefit analysis put together by an internal client, to previous work done on a project by the IT group. The historical information is a starting point for answering questions such as,

- Who are the customers for the product? What are their needs and how will this product benefit them?

- Is the project a subsequent version of an existing product or a totally new product?

- How much time do we have for this project? What is the trade-off between time and value?

- Do similar products exist within our organization? What can we learn from them about what to do or not do?

- Is there competition for this product? Who is the competition and who are its customers?

- What does our organization expect to achieve by completing this project? What is the value of this project?

- Where does the project fit into the "big picture"? Are there dependent projects?

A vision document should contain a short product-capability statement. This vision statement helps team members pass the elevator test—that is, by using the statement, they can explain the purpose of the project within two minutes. As formulated by Geoffrey Moore, the elements of the elevator test are

- *For* (target customer)

- *Who* (statement of the need or opportunity)

- *The* (product name) *is a* (product category)

- *That* (statement of key benefit—that is, compelling reason to buy)

- *Unlike* (primary competitive alternative)

- *Our product* (statement of primary differentiation)

—G. Moore [1991], p. 161.

Although the vision statement for an internal product for customers of IT will vary slightly on the above theme, a vision statement for a fictitious, Web-based collaboration tool might be worded as follows:

> <u>For</u> *Fortune 1000 companies' product development groups <u>who</u> need to build innovative products across both geographic and organizational distances, <u>the</u> Adaptive Software InNovator product <u>is a</u> Web-based, server-layer collaboration tool <u>that</u> speeds turning creative ideas into innovative products. <u>Unlike</u> other collaboration tools, <u>our product</u> integrates accountability into the too often open-ended creative process.*

By answering the questions posed above, the project team puts together a detailed project vision document, which should contain some subset of the following:

- **Project Background**—Describes the current environment(s) that the project will affect directly or indirectly.

- **Project Vision Statement**—Defines the project vision in 25 to 50 words (the elevator test).

- **Project Scope**—Directly sets the boundaries (resource, schedule, scope) on the project so that it can be done successfully. Also specifies what is *not* included within the application, a detail that is useful in understanding the project's boundaries.

- **Executive Sponsor**—Names who has the greatest stake in the project and identifies who has overall responsibility for a commercial product or for an internal project. This person becomes the sponsor responsible for project costs and benefits.

- **Product Market Positioning**—Describes how the marketing department is positioning this product.

- **Internal and External Customers**—Identifies internal and external customers and how they will use the product in their job.

- **Business Functional Objectives**—Addresses the benefits of the product in terms either of opportunities to exploit or of solutions to current problems.

- **Technical Performance Objectives**—Identifies technical performance criteria and measurements.

- **Project Risks**—Describes the major risks that could adversely impact the outcome of the project.

- **Staffing Requirements**—Identifies the skills and number of staff required to develop the product.

- **Prerequisite/Dependent Projects**—Identifies the project's dependencies on deliverables (such as requirements specifications, architectural constraints, or code modules) from other projects.

- **Constraints**—Identifies limits imposed on the project and outside the project team's control, in the form of staff, budgets, interfaces with other systems, technology, or time.

- **Assumptions**—Identifies all costs, benefits, and situations underlying or having a bearing on the project proposal.

The Project Data Sheet

Where do the project team, other stakeholders in the project, or people with a casual interest in the project go to get a thumbnail sketch of the project? The project data sheet! The PDS is the minimum deliverable from any project initiation activity.

Whatever the detailed contents of a project vision paper, a one-page PDS should also be developed. For some projects, the PDS either may be enough by itself or may need only minimal supporting information in order to constitute a complete mission statement. The PDS captures the essential nature of the project in a simple but powerful fashion.

As the quip "I would have written a shorter letter but I didn't have time" demonstrates, condensing an enormous volume of project information into a single page forces the team to carefully consider and select the most important parts of the project. The very act of sifting through and organizing information helps team members focus on important aspects of the project.

The PDS includes the following details:

- clients / customers
- project objective statement
- features
- client benefits
- performance / quality attributes
- architecture
- issues / risks
- major project milestones
- core team members

The project objective statement (POS) should be specific and short (25 words or less), and it should include important scope, schedule, and resource information. Referring to the vision statement created for the InNovator product, the POS might be "Specify, develop, and prepare for market a new Web-based, server-layer collaboration tool called *InNovator,* by the end of July 2001, for an investment of approximately $500,000."

The Product Mission Profile

The ability to create a mission comes from an understanding of the company's strategic focus—from such dimensions as product leadership, operational excellence, or customer intimacy—and from an understanding of the marketing strategy for a particular product. If, according to Michael Treacy and Fred Wiersema (Treacy95), an entire company, even one as large as Hewlett-Packard or IBM, needs a single strategic focus, surely a product team needs a single focus also. Focusing on customer intimacy does not mean ignoring the other two dimensions; it means concentrating on whatever dimension will offer the company the greatest competitive advantage.

Marketing strategy must be considered in the context of product value. Developing software for a piece of in-flight avionics equipment in which a defect could cause serious injury or death, and developing the next Internet browser pose fundamentally different challenges. The market for each dictates a distinct development strategy. Clearly, these are two unique environments. The first case demands overwhelming excellence in the eradication of defects. No one is going to be overly worried about a few months' delay. This does not mean that *any* schedule will work, but that a reasonable, adequate, "good enough" schedule is acceptable. In the second case, however, schedule—and a very fast one—is the driver. Defect levels shouldn't be excessively high, but adequate, reasonable, "good enough" defect rates may be acceptable.

The product mission profile is an important tool for documenting focus—a contract between the development group and the executive sponsor or primary customer. While companies focus on strategy, project teams need to focus on the priorities of major product attributes.

Table 3.1 shows a matrix of attributes that give a product its value—its scope (features), delivery schedule, defect levels, and resources (cost, staff, and equipment). The table displays the relative importance of each dimension and provides focus for the development team. The first two columns, Excel and Improve, can contain only one entry each, but the third column, Accept, contains two entries. If the focus of excellence is to ship a product with world-class features, then everything else takes second place. Schedule might be designated as warranting improved performance, but it would be less important than the product's features. (If the schedule slipped seriously, the schedule itself might assume a temporary position of higher priority, but it still would not be the focus of competitive excellence.)

Table 3.1: Product Mission Profile Matrix.

Product Quality Dimensions	Priority Level		
	Excel	Improve	Accept
Scope (Features)	◆		
Schedule		◆	
Defects			◆
Resources			◆

As shown in the table, the Excel column identifies Scope (Features) as the most important characteristic for marketplace success. Given this matrix, trade-off decisions that the team makes during the project would favor feature richness over development speed. The second column, Improve, indicates the second most important characteristic. Within the constraint specified by the first column, the team tries to improve the indicated characteristic—schedule—as much as possible. In the matrix shown, team members try to improve the schedule unless their action results in unacceptable reductions in features.

An entry in the Accept column indicates that a characteristic can be more variable than the others. If Resources are given an Accept-level priority, a wider variation in staffing or cost would be acceptable. The Accept column indicates the characteristic needs to be "good enough" and is the first considered when any trade-off is required. In the example shown in the table, the team and the product's sponsors have agreed to expend additional *resources* or accept higher *defect* levels (obviously within limits) if necessary to meet schedule and feature requirements. "Good enough" does not give team members carte blanche. An *acceptable* level needs to be defined. For example, the budgeted cost for a project might be $250,000. If cost were indicated as a priority in the Excel column, $250,000 would be a maximum amount and the team would strive to spend less, or at least no more, than this limit. If cost were indicated as an entry in the Accept column, the "good enough" range might be plus or minus $50,000.

Every project team has to adapt to external and internal change during a project's life by making trade-off decisions. These decisions are difficult because of each team member's natural tendency to try to excel in all dimensions. Without explicit priorities, the executive sponsor has to be involved in most change decisions, consuming valuable project time. Explicit trade-off priorities give the team members a basis for understanding how various factors relate to each other so they can make the decisions themselves.

In putting a mission profile matrix together, the team must stay within the bounds of what is reasonable and feasible. Marking a dimension as acceptable and then creating extreme acceptability measures clearly defeats the purpose.

It would be unusual for a complex project to have its highest priority in either the resource or defect dimension. If the focus is on low defect levels, it is improbable the schedule focus could be anything but acceptable. Extremely low defect rates and high speed are generally not compatible. High speed and acceptable defect rates, however, are.

The Product Specification Outline

A short focusing statement such as the project objective statement or vision statement is very helpful to the project team and other involved parties, but it is not sufficient to properly determine scope and size or to understand the product. There must be a level of product specification more detailed than the 25-word project objective statement and yet less detailed than a traditional specification document. This intermediate-level document is called the product specification outline.

The PSO serves several purposes. First, it provides the stakeholders and core project team members with a reasonable understanding of the boundaries and scope of the development effort. It is important to specify what is included and, sometimes even more importantly, *ex*cluded from the product. New projects often suffer from unbounded expectations. Developers and customers see the new product as solving all the old nagging problems. Expectations need to be tempered with a dose of reality.

Second, the PSO is the baseline for size estimation. Whether the team is developing a work-breakdown structure and estimating each feature or using a tool such as function-point estimation, a reasonable project size is necessary for rational project planning. Reasonable sizing requires an understanding of at least an outline level of project requirements.

Third, the PSO facilitates adaptive cycle planning, which is covered in greater detail in the next chapter, and is accomplished by assigning product features to specific cycles (similar to project milestones). In order to develop these plans, the basic features or functions of the product need to be identified. Whereas purely task-based planning can be accomplished (although it is not recommended) by a cursory analysis of the requirements using a task-list template or methodology from prior projects, *adaptive* planning requires knowledge of the features of the application to be built. Outline specifications provide the required information.

The specification outline's primary objective is to define the features or functionality of the product. For the purposes of creating a specification and subsequently for planning iterative cycles, I use the term "component" to indicate a set or group of features. Although similar to the concept of component in object-oriented development approaches, here the term is used to define a group of things (objects, business features, the graphical user interface (GUI), or containers, for example) that are planned and implemented together rather than in the more restrictive sense of components as groups of objects. I use the words "component" and "feature" (a component is a set of features) because Adaptive Software Development does not depend upon, or specify, particular software engineering techniques such as object development, data flow diagrams, or entity relationship data models. Adaptive Software Development is a management approach to delivering software products; it is not a specific development approach.

Three component types are used in ASD—primary, technology, and support. Primary components deliver functionality to the customer. A list management component in a spreadsheet program would contain a number of features to implement that functionality. In a business application, a component contains the functions and data required to implement some business process such as generating orders or producing warehouse stock reports. A primary component could be represented by a process in a data flow diagram or by a use case in an object-oriented analysis document.

Technology components—networks, computer hardware, operating systems, and database management software—are those on which the primary components are built. Often, many of the technology components are already in place and only need to be used by the development team. However, if the technology component is not already installed, it must be identified and installed as one of the project team's responsibilities. For example, if new database software is part of the project, its installation is identified as a technology component.

Support components include everything else, from data models to conversion programs to training materials.

In *Dynamics of Software Development* (McCarthy95), Jim McCarthy provides definitions, paraphrased below, that could be incorporated into a PSO for organizing features:

- *strategic features* are centered around fundamental choices such as operating systems and hardware platforms

- *competitive features* respond to features the competition had or might implement

- *customer satisfaction features* are those frequently requested by customers

- *investment features,* usually of an architectural nature, are those offering long-term benefit, but not much short-term benefit

- *paradigmatic features* are those that change the way people work and therefore have significant competitive implications

Product specification outlines are used in initial estimating and planning, but they may not be detailed enough for development. As each development cycle begins, the team will need to decide how much additional detail is required. In *Rapid Development* (McConnell96), Steve McConnell discusses how specification outlines grow into the *minimal specifications* necessary for the project, a topic germane to the minimalist documentation approach needed for extreme projects.

As McConnell points out, detailed specification writing is often wasted effort because rapid changes occur during the project, making the specs obsolete. However, there are also potential problems with minimal specifications, particularly if the team uses the minimalist concept as an excuse to avoid a seemingly tedious activity. Developing specifications is extremely important, but the specs do not need to fill thousands of pages in order to be effective. Writing a useful spec is another one of the balancing acts required for extreme projects.

Share Mission Values

Documenting mission artifacts is a mechanical task. Creating the shared values that infuse meaning and purpose into the artifacts requires interpersonal effort. Sharing a mission—understanding the subtleties, agreeing on meaning, generating passion—doesn't happen in a two-hour meeting. With the team's concerted effort and patience, shared values grow over time. The core team needs to be aligned, but other stakeholders at various levels of the organization also need to be brought in sync—a formidable task, for a mission is like a magnetic field, with alignment based on its strength of shared values, as depicted by the three variations in Fig. 3.2.

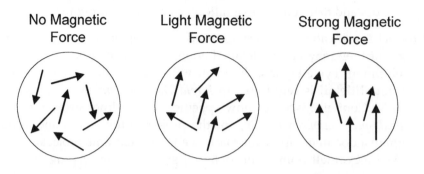

Figure 3.2: Shared Mission Values Show a Magnetic Alignment.

Too many managers and staff think *sharing a mission*—aligning values—means getting together for a couple of hours and jointly writing a mission statement on a flip chart. Unfortunately, sharing a mission is much more complicated than merely writing down a list of values. Its difficulty can be likened to the experience of veteran rock climbers who are always admonishing less experienced climbers to *use their feet*. The reminder is a kind of miniature mission statement. The intent—to get the climber to use leg and core body muscles rather than arm muscles—is correct, but hearing it and understanding it are miles apart. In rock climbing, mastering very precise foot placement, initiating body movement from the legs, and directing momentum through the hips rather than pulling with the arms takes hours and hours and hours of practice. Getting a team to share a mission is equally challenging.

How does an $8-billion company turn virtually on a dime? Of course! Its management writes a new mission statement and off it goes in a new direction! As absurd as it sounds, that is just what Microsoft did in early 1996 (Rebello96). Within a six-month period, the Internet went from being a sideline to the main line at Microsoft. With most 15,000-employee companies, such a radical change would have taken much, much longer—if it happened at all.

What makes Microsoft different? First and foremost is the Microsoft employee's trust in Bill Gates and others in leadership positions—a trust that the mission moves the organization in the right direction, and that, while difficult, it is still achievable. There is a direct correlation between trust in one's management and commitment to a mission.

A second condition present at Microsoft in 1996 was the staff's sense of shared purpose or values. If Gates had outlined a mission to move into the hog-feeding business, the staff's response might have been different. Good mission statements need to tap into the broader corporation's shared values as well as those of the development team.

Teams don't jell overnight—they jell over a period of time. During the process, each team member must ask questions such as

- Does this project's mission connect to something I consider purposeful?

- Do I understand what to do?

- Do I trust the leadership and the other team members enough to collaborate, rather than always insist on "my way"?

- Can I develop a sense of excitement about and commitment to this project?

Not every project manager has the leadership credentials of a Bill Gates. But there are certain attributes that bring a mission to life for a project team:

- The core team works on developing the mission components together. Refining the mission, or reshaping people's perception of the mission, is viewed as a *continuous* activity, not just a checklist item at the beginning.

- The project data sheet is prominently displayed on wall charts in the team area along with documents related to the project. The display of important project information is intended as a reminder, not as a slogan. While slogans attempt to convince the audience of something, the wall charts help remind it of a vision already agreed to.

- When decisions must be reached or controversies need resolution, the mission components are consulted. In the heat of a project, controversy can escalate. Often, the resolution begins by reviewing the mission statement in order to identify an area of agreement that can be used to initiate more constructive discussion.

Quality

Software development teams cannot successfully share values related to mission statements without discussing quality. Few individuals strive to accomplish poor-quality work, yet most individuals are frustrated by the quality of products produced by their organizations. It seems there is always *someone else* whose work degrades *our* quality.

Despite all the words written and delivered in seminars and at conferences about the issue of assuring quality, there are still significant problems in practice, such as the following:

- failure to differentiate among perfection, excellence, and "good enough"

- failure to acknowledge that defects are only one aspect of a product's value

- failure to develop a shared understanding of a particular product's expected quality

- failure to understand that quality is an emergent property that is often ambiguous and uncertain

- failure to understand the unique characteristics of software that negate application of practices used in other functional areas such as manufacturing

For something so deeply valued in our software development culture, quality remains hard to define. For me, the most useful definition of quality is the following:

> *Quality is value to some person.* —G. Weinberg [1992], p. 7.

The simplicity of Weinberg's definition belies its depth. Two critical issues are suggested. The first is that quality is multidimensional. Many software engineers consider defect levels to be the sole dimension on which to measure quality, but Weinberg's use of the word *value* indicates a definition of quality that is broader, to be measured in terms of schedule, scope, defects, or even aesthetics. A second issue raised by the definition is whether quality is an intrinsic characteristic or is dependent on a viewer's perception of the product. Too many definitions of quality leave out consideration of how people are affected by the product and, therefore, are sterile and lifeless.

Many software engineers are unsettled by the idea of value as a perception of the beholder. Their analytical nature wants quality to be intrinsic. One of the most telling arguments for quality as a *perceived* property is the deep emotion it stirs. By defining quality in an analytical way, by reducing it to charts and numbers and processes and rules, we lose passion—the very motivator needed to accomplish it.

The current state of software quality management practices often is governed by the phrase, *Do it right the first time*. But in a complex environment, *Do it right the first time* is a recipe for failure. In essence, it says,

- We can't be uncertain.

- We can't experiment.

- We can't learn from mistakes (because we shouldn't make any).

- We can't deviate from plan.

In the early stages of a project, if the delivery time horizon isn't too far out, we may be able to speculate on what the *generally correct* direction is, but defining a single "right" borders on fantasy. Even if we *could* define right, doing it the first time makes no sense for anything other than trivial products. "The first time" assumes we understand the cause and effect, the specific algorithm of getting to the final product from our initial starting position, and the needs of all stakeholders. It says we know it all. What we need is to be willing to *do it wrong the first time* in order to *get it right the last time*.

Several writers have addressed this multidimensional view of quality, most notably James Bach. Bach's terminology set off a minifirestorm of reaction: He used the term *good-enough software development* as a tag line (Bach95). The irony is that "good enough" seems to indicate a compromise position, one of settling for less than the best. It offends many developers whose value systems seek perfection rather than quality.

To me, "good enough" means the combination of attributes providing the *best total value* in a complex environment. With the myriad combinations and permutations of value attributes—scope, schedule, defect levels, and resources—there can never be an optimum value. Not only is the value landscape vast, but competitors are constantly altering its features. "Good enough" suggests synthesis rather than compromise. It does not mean settling for average, but advocates

"The rule 'I must be perfect' is common enough in the general population, but among people attracted to software engineering, it's close to universal."
—G. Weinberg [1994], p. 212.

delivering the best mix of attributes in a given competitive situation. If software development is to be considered a true engineering discipline, then balancing—not compromising—conflicting quality values is part of building high-quality products.

Value is related to use. For example, with a piece of software slated to run medical equipment that could injure or kill a patient, the *value* goal would be *low defects,* no matter how much time the development and test phases take. However, the same value standard would not be applied to a sales application needed for product planning.

The four broad categories of software quality in Fig. 3.3 are the same ones used in the product mission profile matrix (Table 3.1): scope, schedule, resources, and defects. Many project management approaches use only the first three categories, but I include defects as a major category because of the emotional reaction many people have to the issue of quality and the tendency to consider defects to be synonymous with a lack of quality.

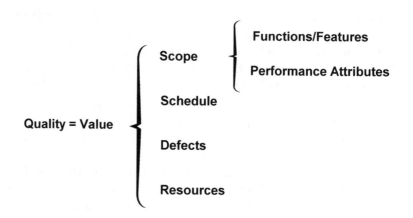

Figure 3.3: Dimensions of Quality.

The belief that a product with zero defects can be achieved is a myth left over from the manufacturing roots of most quality assurance initiatives. In manufacturing a thousand widgets per hour to a statistical quality tolerance of $+/-.001$ inches, a plant that can manufacture so that no widget exceeds the tolerance would be considered to have a zero-defect level. Would the narrow tolerance have any real meaning if the best machine tools in existence could only produce to $+/-.1$

inches? Of course not, yet we openly embrace equivalent nonsense when it comes to software.

Depending on what they are designed to do, software products have different tolerances for defects. The tolerances for an operating system, a sales forecasting system, a medical equipment control system, and a 3-D adventure game are very different. Many organizations resist this idea of *tolerance* for software defects, and have little clue as to how many defects their software actually contains. It is interesting to me that the better software developers usually know more about their defects and which ones they are shipping to customers than the developers (and companies) who claim a belief in zero defects know about the defects they ship.

Evaluate the Mission Every Day

Creating a sense of shared values, a sense of a shared mission, requires constant attention by the team members. In the early stages of the project, there may be broad ideas about the feature set, but the details are off in the fog. I once ran a six-month, twenty-plus-person project with a reasonably stated overall mission, but fuzzy details. The project plan contained about eighteen tasks, which we never looked at after the first week. But every single day, in some fashion, the team discussed the end goals, the contents of the final product, and what the product's features might look like. Each team member had "to-do" lists according to his or her own responsibilities but, from an overall management perspective, meeting the cycle dates and refining the component definitions were the main focal points.

In complex projects, no one *completely* understands the final deliverables until they are ready for shipment or installation. The question in everyone's mind, every day, is, How can we improve our understanding of what we are supposed to produce? Shared understanding itself is not a goal, but an ongoing process of interaction and relationship-building.

Focus on Results

The four steps that define the recipe for success on a complex project are: Define the result; Define the quality criteria to measure the result; Create a shared understanding of the destination; and Let the results

*"**Great Groups** ship. Successful collaborations are dreams with deadlines."*
—W. Bennis and P. Biederman [1997], p. 214

evolve in an emergent fashion. A feature team of six people who have their own ideas about what constitutes quality and who are driven to perfection is a recipe for disaster. The same group, driven by a vision and guided by a focus on excellence—that is, a group that understands the pitfalls of perfection—will still be witness to colossal arguments and bruised feelings, but will stand a better chance of shipping a great product. Great groups, and particularly *leaders* of great groups, understand the difference between perfect, excellent, and "good enough."

One of the critical success factors in mountaineering is the climber's ability to focus intensely—on the next step, then on the next, and so on. I've been in situations in which I had to talk myself into pushing ahead. Counting each crunch in the glacier ice, I more than once have felt overwhelmed by the need to take even one hundred steps. Reaching one hundred, my next goal was another hundred—thinking about getting all the way to the summit was just too daunting. By playing games with my mind, I could fool myself into taking the climb step by step. In climbing, just as in product development, you know one hundred steps isn't the real goal, but it provides relief—a small accomplishment to battle the fatigue and the little voice saying, "Go back; you've done enough. It was a valiant effort. The top isn't really so important anyway." Good mission statements help team members battle the doubt, indecision, and despair that is part of any project.

In the beginning, a mission statement is purposefully broad, but as the project's ship-date approaches, mutual understanding narrows the gap so a product can be shipped. A mission is like a mountain, broad at the start, increasingly narrow at the top. The more extreme the software project, the more crisply defined and visual the focal point must be as the end approaches. Head-down, fur-flying ship-mode is not the time to be uncertain about the goal.

Quality is not about time—software archaeology is littered with the bones of very long projects of very poor quality. Quality is about setting the correct mission and the correct objectives. It is about understanding the constraints, managing trade-offs, and displaying courage. If the objective is to produce a certain feature set in three months, and there are no specified criteria for the software to integrate with other applications, then labeling the result as poor quality because of poor integration is spurious. If the plan is to monitor quality through the use of technical reviews and they are bypassed because of time constraints, then someone has either made a valid trade-off decision or lacked the courage to take the time needed.

Much of the outcry about poor quality is based on the deterministic view that quality can be predicted and planned, and that specific processes can be put in place to achieve it. In reality, quality is an emergent property, an ever-shifting position on a fitness landscape. Until we learn to view quality not as a point in quality space but as shifts along a continuum of unfolding possibilities, the arguments will continue to focus entirely on the wrong issues.

Summary

> Thomas Jefferson's mission statement to the Lewis and Clark expedition was one of the most famous in history. It combined a specific goal, boundaries of behavior, and wide latitude for implementation.

> A mission statement needs to be focused. Attempting to excel in multiple dimensions usually results in a product being mediocre in all of them and excellent in none.

> The product mission profile forces a focus on the single area—features, schedule, defects, or resources—in which the development team needs to excel. This profile provides the high-level, trade-off strategy for the project.

> A mission statement facilitates collection of information that is relevant to the project's desired result.

> A mission should establish direction, inspire the participants, and provide enough detail for ongoing decision-making.

> The components of mission are a project vision (or charter), a project data sheet, and a product specification outline.

> Writing a mission statement is easy. Creating a sense of shared responsibility for achieving the mission is very difficult. Building a shared vision is an ongoing, never-ending, collaborative team effort.

> The ability to periodically de-focus is important.

➢ The project vision or charter establishes the focus and key motivational theme for the project. It establishes which direction to take into the fog of the unknown. It provides the boundaries for the exploration phase of the Speculate–Collaborate–Learn life cycle.

➢ The project data sheet is a one-page summary of the key information about the project. It serves as a focal point and quick reminder of the most important elements about the project. It is simple, but powerful.

➢ The product specification outline describes the features of the product in enough detail so that developers can understand the scope of the effort, create a more detailed adaptive cycle plan, and estimate the general magnitude of the development effort.

➢ Quality characteristics are part of the mission definition. Software developers often fail to differentiate between excellence and perfection. They also fail to provide a basis for making necessary trade-offs during a project.

➢ Quality is in the eye of the beholders. In Jerry Weinberg's words, "Quality is value to some person."

CHAPTER 4
Planning Adaptive Development Cycles

Complex projects force us to undertake activities that are more difficult than ones we have successfully completed before. On such projects, we must attempt the impossible, or at least the nearly impossible. Those who aspire to be successful with complex projects must be able to manage risk and tolerate change. While climbing involves more risk than a Sunday morning golf outing, it is less risky than non-climbers perceive. Climbers must learn to manage risk and tolerate environmental change, or they become ex-climbers.

Because climbers routinely attempt the unattainable, they have developed techniques to increase the likelihood of success. The technique rock climbers have perfected is called *redpointing*.

Unlike traditional on-site climbing, which has the climber starting at the bottom and climbing to the top, periodically placing protective anchors along the way, and attempting climbs only incrementally more difficult than ones previously completed, redpointing involves having the climber learn how to execute a climb through repeated ascents of the same terrain. The technique allows the climber to practice difficult moves and provides an opportunity to make mistakes and learn from them. Eventually, the climber must complete the entire climb—just as in an on-site ascent—but the difference is in the repetition of moves.

Redpointing conditions the climber, who practices a move, or a series of moves, over and over—until the body *knows*. It provides a technique for learning what Dale Goddard and Udo Neumann call *motor engrams*. As defined in *Performance Rock Climbing*,

> *Engrams are complete records of movements we've made. More important, they're also the instruction manuals for reproducing the same movements at will. Engrams contain all the directions necessary to reproduce living versions of moves stored in memory. You could think of an engram as a packet of preset muscle instructions on how to reproduce a particular move.*
> —D. Goddard and U. Neumann [1993], p. 15.

Climbing above one's limit requires knowing the sequence of every handhold and every foothold, how much effort to use on every hold, how much body tension to exert, each subtle weight shift—and more. This type of learning is only accomplished by repetition, and can take days, weeks, or even years. In fact, Goddard worked on redpointing a climb at Smith Rocks in Oregon, the hardest climb in the country, off and on for three years.

Redpointing makes the seemingly impossible possible. Typically, climbers can redpoint climbs one to two grades harder than they can accomplish on a traditional, bottom-to-top, on-site climb. By dividing a climb into small sections and repeating moves until they are familiar with them, climbers gain enough knowledge and confidence to complete the climb.

Adaptive development provides a similar technique for approaching very difficult software projects. Adaptive techniques stimulate learning *mind engrams*, enabling the software engineer to build the knowledge necessary to accomplish complex projects in bite-size, iterative chunks. Traditional software development techniques are analogous to those used in on-site climbing: They leave developers without the benefit of iterative learning. Adaptive practices allow developers to advance or to recede; literally, to back up to find an alternative path. Traditional techniques work as long as the task to be accomplished draws on the developer's realm of experience, but they don't allow for the intense learning required to redpoint a tough project that challenges one beyond the limits of one's experience. Iteration and learning, as we have seen within the context of redpointing, give the developer a better chance for success.

This chapter presents specific adaptive development planning techniques. Adaptive development proceeds in cycles or iterations during periods of time in which both new components and suggested modifications to previously completed components are constructed. The five key characteristics of adaptive development cycles are defined below.

First, adaptive cycles are *mission-driven*. The mission artifacts identified in Chapter 3 drive the development effort. Second, the cycles are *component-based* rather than task-based. The focus is on refining the definition of the components—the desired results—not on listing the myriad activities necessary to produce the results. Third, adaptive cycles are *iterative*. Results are achieved by conducting successive iterations in which the software product converges on an implementation of the desired end-result. Fourth, the cycles are *time-boxed*. Both the entire project and each individual development cycle are confined to specific time periods. Complexity and uncertainty can lead to indecision and confusion. Time-boxing is a mechanism for forcing hard trade-off decisions and producing tangible results. The fifth and final characteristic is that adaptive cycles are *risk-driven and change-tolerant*. Risk is the probability of an adverse event impacting a project. Some risks can be mitigated, others occur and cause actual changes. Change-tolerant processes transform changes into an opportunity to gain advantage over a competitor's less change-tolerant approach.

Characteristics of Adaptive Cycles

Figure 4.1 shows details of the various stages—also called phases—of the Adaptive Development Life Cycle first introduced in Fig. 2.3. The Speculate stage is divided into a project initiation step—where basic project mission statements, project management information, and initial requirements are developed—and an adaptive cycle planning step—where components are assigned for construction or assembly. The Collaborate stage is divided into cycles, each of which delivers components to the customer. The Learn stage involves the examination of the product from customer- and technical-quality perspectives and also reviews the project team's performance. The loop arrow shown in the figure would typically be labeled a feedback loop. But feedback implies matching a monitored signal with a reference point and making necessary corrections. A learning loop is more than that because it involves people, the uncertainties of the reference point (the plan), and the difficulties in measuring results. A feedback loop implies simple signals. A learning loop implies complex human interactions. The learning loop is adaptive development's redpointing technique.

Before presenting the steps in adaptive cycle planning, we need to explore in more depth the five characteristics introduced in the preceding section.

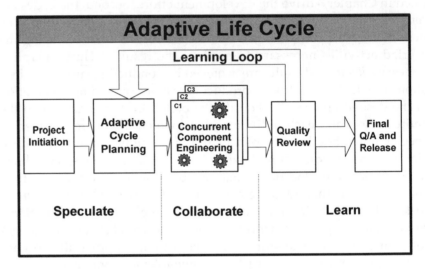

Figure 4.1: The Detailed Adaptive Life Cycle.

Adaptive Cycles Are Mission-Driven

As explained in Chapter 3, success in adaptive development depends on project team members having a good definition of the project's mission and a solid sense of direction and purpose in order to focus development. Adaptive cycles enable the project team to explore the mission in more detail and to make adjustments as the project proceeds. Adaptive cycles provide a structure that encourages, even demands, that the team view the project as a learning journey.

Adaptive Cycles Are Component-Based

In the previous chapter, a component was defined as a group of things, (for example, product features) that are planned and implemented together. Planning an adaptive project is very different from planning one based on a traditional life-cycle approach. Traditional software project management is primarily task-oriented, whereas adaptive management is result- or component-oriented first, and task-oriented second—a two-dimensional planning approach. Task-oriented development leads to task-bloat—that is, the planning of tasks to the finest level of detail. For example, the specification process can be divided into an endless string of tasks, from data modeling to the gathering of security specifications. A task-oriented focus may not produce a good project plan primarily because any competent developer or analyst can spew out a long list of required tasks without knowing much at all about the product under development. Task lists from the last project are copied, the hours tweaked, and *bingo!*, the project plan is done. The problem is that few project team members understand what is to be delivered, and the "plan" is not a plan at all.

what we did!

Evidence of task-bloat abounds. During a recent consulting assignment, a project manager asked me to look at his plan. Without opening the document, I said I understood the problem. In response to his somewhat incredulous expression, I explained that he was using his work-breakdown structure as a "to-do" list (I could tell by the number of pages). Sure enough, the plan was laced with one-, two-, and four-hour tasks! Tasks are important, but components, and the features that make up the components, are more important. In projects in which component-planning precedes task-planning, the team members invariably have a clearer understanding of their objectives. The process of identifying components and allocating them to cycles gives the team a better picture of the product and the project than it can get from a purely task-oriented approach.

Adaptive Cycles Are Iterative

Most writers and consultants on software development practices advocate some type of iterative development. Even the military is abandoning the waterfall model (Sorensen95). However, when the project goal is certain and the environment is stable, a serial waterfall model may still be the most efficient approach. As the goal becomes less certain, and the number of changes to product requirements increases, success in development depends more on continuous learning about the product than on predicting its final form.

In the traditional waterfall model, water falls; it doesn't rise. Everyone involved in software development has seen the traditional waterfall picture—with boxes and arrows flowing from the upper left of a page to the lower right, and with smaller arrows cycling back to the previous step. Good waterfall development incorporates feedback; it's not as if the developers expect everything to happen correctly the first time. Unfortunately, such approaches expect *most* things to happen correctly the first time and so the primary difference between waterfall and adaptive models is that the adaptive model focuses on *cycling* while the waterfall model focuses on *flow*.

Inarguably, waterfall models recycle and adaptive models flow, but the significant difference between the models is in the primary focus. The argument is reminiscent of the process-versus-results debate. Both are needed to succeed, but whether one views process as driving results or results as driving process provides the viewers with different mental models of the world. This change in perspective represents a significant mental-model change for most people and organizations. It means that the *redoing* is more critical than the *doing*.

This dichotomy between a doing and a redoing mental model is evident in how people approach certain activities. For example, in waterfall projects, change requests are viewed as negative occurrences—as an indication that something was done incorrectly or that something was missed. In an adaptive project, change requests are viewed positively—as opportunities for learning and as a means of attaining competitive advantage.

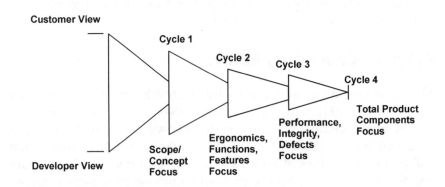

Figure 4.2: Converging Views in Adaptive Cycles.

Figure 4.2 shows a project that begins with a divergence between the customer's and the developer's views of the product. During each successive cycle, the stakeholders learn more about each other's view of the feature sets. The figure illustrates that as this learning takes place, the differences narrow until the final product emerges to satisfy both the customer's and developer's requirements.

While many people see the benefits of a cycling approach, there is usually a fear that it will involve endless iteration without progress. Developers complain about customers whose requests make complete flip-flops from session to session. They cite customers who change their requirements on a daily or weekly basis, forcing iteration without progress. Such problems can afflict the technical side of the team also. Developers may find themselves continuously changing designs or constantly revising models. For Adaptive Software Development to be successful, everyone involved must understand the difference between healthy, iterative focusing and de-focusing and unhealthy oscillating.

We have learned that keeping adaptive projects on track means paying constant attention to whether the project is cycling or oscillating, but how can we differentiate? A simple answer is that cycling projects focus and de-focus, but ultimately converge on a solution. Oscillating projects generally diverge from a solution. The team, and especially the project manager, must be constantly attuned to whether true progress is being made, although this can be difficult to assess in the short term. Oscillating projects usually have two main sources of problems: the project mission and the problem definition. First, some team members may not understand, or may not agree with, the stated objectives of the project. The essential action to take when oscillation occurs is to revisit the project's mission statement and confirm, Is this our objective? and Is this still our profile matrix? Often, this re-grounding is all the team needs to get back on track.

Second, oscillation may occur if either the customers or the developers have a faulty understanding of some problem that arises during the project. A customer who continually changes his or her mind about a feature or a developer who continually re-codes a particular module is usually signaling a lack of understanding of what is expected. At this point, the project team needs to back up and redefine the problem.

A final point about the nature of iterative, adaptive cycles is that they promote the concurrent development of components. Short development times demand concurrency. Adaptive cycles keep concurrency under control.

Adaptive Cycles Are Time-Boxed

It took me several years of using time-boxing techniques on client RAD projects to come to the startling realization that time-boxing is not really about time—it is about forcing trade-offs and learning in the context of short delivery cycles. At the most basic level, time-boxing is the process of setting specific due dates both for the project and for each individual adaptive cycle. A specific due date provides a clear focal point, which contrasts with the ambiguity and uncertainty of a complex project. However, fixed dates can be used inappropriately as a bludgeon to coerce team members into working overtime and cutting corners on quality.

A time-box should be used to force hard trade-off decisions, not to pressure team members into making poor decisions. Trade-offs encourage all parties—managers, customers, and developers—to make good business and technical decisions. Time-boxes force these decisions throughout the project. Without this continuous pressure to make hard decisions, human tendencies push them to the end of the project when options are usually very limited.

Imagine, for example, that at the end of the second development cycle of a six-month project, the team realizes that the planned functionality of the product cannot be accomplished in the time frame allocated. Further analysis, based on better information from three months of working on the project, indicates that all the functionality can be delivered in a total of seven months. The team, in conjunction with its customers, can then make an informed decision about whether to meet the due date by cutting functionality or vice versa. Without a periodic, realistic evaluation process, the development team may make the wrong decision, or even worse, postpone the decision too long.

The second key benefit of short delivery cycles is that they force convergence and learning. Every team member wants to produce high-quality work and, therefore, is reluctant to submit any partially completed component for peer review. Our mental model, which derives from our inner sense of quality, associates half-completed work with shoddy work. Instead, we want team members to think that half-completed work is just that—half-completed work—something we can share with others and learn from so that we don't inadvertently waste time. Time-boxes help overcome our natural tendencies to overwork a deliverable before we ask others to review it. Time-boxes provide a gentle, but firm nudge to learning.

A third benefit of time-boxing is that it helps keep team members focused. Having to produce a portion of the product and present it for customer review every few weeks or months provides an incentive for developers to concentrate on the most important activities.

portion of a product

Adaptive Cycles Are Risk-Driven and Change-Tolerant

Traditional project management focuses first on detailed task-planning and then on monitoring each task's deliverable in the context of its pre-determined requirement. Whether explicitly articulated or not, the premise of traditional project management practices is that most activities will proceed as planned and only a few will be subject to change. Teams using adaptive practices assume that many aspects of a project will change during the development. An adaptive team is more concerned with responding to, rather than controlling, change.

docs → pubs → training

Creating change-tolerant software—that is, software that is maintainable, enhanceable, and modifiable—is a goal of Adaptive Software Development. (See Charette89 for more on risk-driven and change-tolerant systems.) Creating change-tolerant software in a complex environment requires a change-tolerant development process. Development teams must constantly ask questions such as, What could change? How would we respond to that change? How can we mitigate the impact of different types of changes?

Unrestrained change creates chaos. The nature of adaptive cycles both promotes change-tolerant activity and places boundaries (in the form of shortened time frames) to help manage change. A product's position on its fitness landscape defines its relative success compared to its competitors. In rough fitness terrain, not only is it difficult to ascend higher peaks, but competitors alter the landscape all the time. Change-tolerant processes necessitate a constant scanning of the environment, modifying products to respond to competitors' moves, and creating change to which competitors must respond.

Risk measures the probability that a development effort will not produce a viable product. Demonstrating a project's viability involves, in part, understanding where the probability of failure is highest and developing strategies either to reduce the probability of failure or to accelerate the project's demise. For example, if one risk to a particular project is that we are using a new untried technology, then we should assume that the technology may fail in this instance. A very poor strategy would be to spend the first development cycle creating comprehensive use-case models. To do so, the team would have had to spend considerable resources, but would be no closer to answering the most critical viability question for the project. Now, the team must spend even more money to come to the conclusion that the technology will not work. In planning an adaptive project, developers will find that it is usually a better strategy to place high-risk items in early cycles.

Adaptive Planning Techniques

The planning practices presented in this chapter are geared to small to medium-size projects. This basic Adaptive Life Cycle provides enough fundamental structure to use on projects ranging up to 10,000 function points, although in complex environments, factors other than size determine the need for additional rigor. In Chapter 9, an Advanced Adaptive Life Cycle is presented, which provides techniques for scaling adaptive development up to much larger projects.

While planning in a complex, turbulent environment may be speculative in nature, the project team must have a good road map. Although far from ideal, a half-right, good-enough plan is still better than no plan. The wisdom of this statement rang especially true on one recent pilot project in which a client project manager commented on adaptive cycle planning:

> You know, most of the things we've done in this adaptive project we've done before in some fashion on other client/server projects. One of the things that is very different, however, is planning which features go into each cycle. In my last project, which lasted about six months, incorporating this planning technique would have cut two months off our six-month schedule. We wasted a lot of time in iterative, prototyping activities, without having a very good game plan.

In the remainder of this chapter are techniques for planning adaptive cycles and then application of those techniques to a hypothetical, but typical project.

Defining Versions, Cycles, and Builds

There are three types of iterative loops important to adaptive development. A *version*, which is the longest loop, produces a new rendition of a product. *Cycles* are major loops used during a project by developers to learn about the product and to monitor the project. *Builds* are the very short loops used by the development team to produce an interim deliverable.

A version is a product that is ready to be installed or, as in the case of shrink-wrapped software, to be manufactured. "Version" as used here, is compatible with the industry standard—that is, whole numbers indicate major releases of a product and point versions (version 4.1, for example) indicate less significant releases. Versions can require time frames of months to years, depending on the project's size.

A cycle delivers a demonstrable portion, or component, of the product to a review process. The most important reason for using cycles is that they make the product visible to the customer. The end of a cycle represents a major project milestone and then provides the opportunity for customer, management, and developer reviews of the interim results. Every cycle has a theme, or a stated goal, manifested in a cycle objective statement. Like a project objective statement, this is a high-level focusing tool for the team. Developers must continually ask themselves, Does what I'm working on contribute to the objective of this cycle? With short cycle times and specific deliverables, project teams are less likely to get off track.

A build constructs an interim portion of the product and makes the product visible to the development team. Depending on the development environment, the overall size of the product, and the cycle involved, the build may be the entire product or only a portion of the product. A build makes the product *public*. A developer may code a portion of a feature or module, but not release it to independent review. A build forces development teams to confront key issues by enabling developers to demonstrate their wares to everyone involved.

"The regular build is the single most reliable indicator that a team is functional and a product is being developed."
—J. McCarthy [1995], p. 111.

Cycle Planning Steps

In an adaptive project, the team must first define the product components, then allocate those components to a development cycle, and finally plan the activities necessary for each cycle. The following paragraphs describe techniques for approaching this two-dimensional (component and task) planning. Steps in adaptive cycle planning are

- Conduct the project initiation phase.

- Determine the project time-box.

- Determine the optimal number of cycles and the time-box for each.

- Write an objective statement for each cycle.

- Assign primary components to cycles.

- Assign technology and support components to cycles.

- Develop a project task list.

Figure 4.3 provides an overview of this approach to planning a project.

Inputs

Cycle Objective Statement

Project Data Sheet

Component List

Risk Analysis

Resource Availability

Deliverable Components	C1	C2	C3	C4
Cycle Delivery Dates	1-Jun	1-Jul	1-Aug	1-Sep
Primary Components				
Order Entry and Edit	x	x		
Order Pricing		x		
Inventory Picking		x	x	
Order Shipping			x	
Reorder Calculation			x	
System Interfaces			x	
Pricing Error Handling			x	
Security Access				x
Technology Components				
Visual Basic	x			
Communications			x	
Support Components		x		
Architecture Document		x		
Conversion Plan Doc.			x	
Data Model Document	x			

Figure 4.3: Component-Oriented Planning.

Step 1: Conduct the Project Initiation Phase

The project initiation phase is a critical period. If the project does not start off well, it may flounder, sometimes for months. The objective of the project initiation phase is to clearly establish project expectations among all the project's stakeholders. The project initiation phase is the stage at which mission statements (defined in Chapter 3) would be produced. The project initiation period should establish the following:

- For the executive sponsor, it establishes the intent of the project and a firm idea of its scope, intended schedule, and projected resource utilization.

- For the customer, it identifies business functions and data to be automated, and the cost and benefits of accomplishing this automation.

- For the developer, it provides an understanding of the product to be delivered and the business reasons behind the project.

The following conditions define the context for the project used as a sample in the remainder of this chapter:

- A feasibility study has been completed.

- The product mission profile for this project is schedule-driven.

- The basic architectural and development technologies have been selected or defined.

- The project size is in the 1,000-to-1,500 function-point range.

- The core team size is in the four-to-eight-person range, and key team members have been identified.

- The project is expected to take approximately six months.

- The executive sponsor has been identified.

- The required training has been conducted.

An adaptive project that has the context just defined can be launched during an intensive project kick-off week. The result of this week-long effort should be a completed project plan (including the adaptive cycle plan) and the product specification outline. These documents would be developed during a series of team meetings and JAD sessions.

The kick-off week should begin and end with sessions attended by the executive sponsor, core team members, and other stakeholders in the project. The executive sponsor should provide an introduction at the JAD session to ensure that team members understand the objectives and business reasons for the project. Having the sponsor take the time to personally introduce the project and convey his or her commitment to the project's success is an important motivational factor. If the sponsor doesn't have time to attend the kick-off week, it indicates that he or she won't be available later to support the project team. At the end of the week, the core team presents an overview of the week's deliverables and obtains the executive's approval to proceed.

A successful kick-off week requires careful planning, which is usually accomplished by a combined effort of the project leader, the JAD facilitator, and selected managers in the client and software development departments.

Step 2: Determine the Project Time-Box

"Never allow team members to jeopardize the product in the attempt to hit what might be, after all, an arbitrary deadline."
—S. Maguire [1994], p. 98.

The project time-box specifies the estimated duration of the project. Project effort and estimates for its duration should be based on industry best practices, such as function-point analysis; however, given the uncertainty of complex projects, estimates may not prove to be particularly accurate. The project time-box should be viewed as a boundary, not as a goal. As the time frame of complex projects extends beyond six-to-twelve months, the probability of any estimate being accurate drops precipitously. Time boundaries help manage the project by forcing trade-offs and by providing review points for learning.

A project's time-box should be both aggressive and achievable. Arbitrary deadlines serve no useful purpose within the development effort. However, a schedule *target date* is a legitimate date that is established by management. A target date may be completely reasonable from a business perspective and, at the same time, arbitrary from a development perspective, so it is important not to confuse target and planned dates. The two dates can be reconciled through negotiations and adjustments to project requirements or constraints, or the two different dates can be maintained and used for different purposes. The important point is that arbitrary dates, which the development team considers both unachievable and irrational, do more harm than good.

Step 3: Determine the Optimal Number of Cycles and the Time-Box for Each

The time allotted a cycle should be on the order of four-to-eight weeks for projects of less than nine months and six-to-ten weeks for projects longer than nine months. Factors such as whether the software is for internal use or for sale also impact cycle lengths. Early cycles should usually be shorter than later cycles to encourage customer involvement. Feedback from focus groups strongly indicates that customers are more impressed with regularly delivered results than with feature depth. Once customers understand that the development team will produce a meaningful product version on a regular basis, the time between cycles can increase somewhat.

A six-month project might have from four-to-six cycles, depending on the nature of the project. The first one or two cycles should be short (lasting four-to-five weeks) to insure engaging the customer early, to verify the project's scope, and to confirm the project's viability. Later cycles (those dedicated to perfecting performance, extending features, removing defects, and preparing for delivery) can go longer, lasting up to eight weeks. Once these cycle durations have been determined, the core team should establish the dates for customer focus-group reviews and confirm them with customer representatives and the executive sponsor. Dates can be adjusted for schedule conflicts, but they should not be permitted to slide more than a week in either direction.

A final but very important point about cycle duration is that it should reflect the development team's perception of the uncertainty about the product's requirements. Shorter cycles should be used for areas of high uncertainty; longer cycles should be used for areas of greater certainty. A single cycle, essentially one using a waterfall approach, is appropriate where the environment and the project's requirements are relatively stable.

For example, a client of mine had a project that involved four major business processes. One of these processes involved a target user group that had little existing automated support, and that functioned in an area proven difficult to automate in other companies. We pulled this particular business process out of the regular cycle plan and formulated a series of short, simple automation steps in order to assess, early on, the viability of automating this process. We believed two to four of these short exploration cycles would be necessary before there would be enough information for us to adequately plan the full development effort. These exploration cycles, in essence, served as an extended feasibility study, producing a trial application rather than stacks of documentation.

The key questions to consider when determining how long a planning cycle should be are, Does it maintain a reasonably short customer-feedback loop? Does it allow enough time to develop a meaningful feature set? Does it provide for each major feature to have at least one cycle available during which focus-group changes can be implemented?

Step 4: Write an Objective Statement for Each Cycle

Just as a project needs a good objective statement to keep the team focused, each cycle needs its own focusing mechanism. As is suggested in the sidebar quote, lack of a theme (an objective) while pursuing a milestone (a cycle) can cause floundering. A detailed task list is no substitute for a clear, focusing statement—a cycle objective statement (similar to the project objective statement introduced in Chapter 3). The detailed task list just insures that everyone will be busy, not necessarily productive. The cycle objective statement provides the next level of definition of the project's mission. It helps the project team solidify decision-making as changes occur during a cycle. Using the project objective statement, component list, and risk-analysis documents produced in the project initiation phase, the team drafts a cycle objective statement for each cycle. As developed in this chapter, our hypothetical cycle plan assumes the following general conditions:

- Customer involvement is hard to establish, and is often perfunctory when achieved.

- Scope issues are difficult to understand and resolve.

- Technical issues are less important to the customer than business issues.

In this hypothetical situation, the cycle-1 objective might be stated as

> **Ensure overall project viability by actively engaging the customer in the project while confirming the feature scope, schedule, and resource estimates.**

Note, until cycle 1 has been completed, nothing concrete will have been produced. Project plans and initial requirements documents are necessary, but they are insufficient to confirm that the project is actually viable. By demonstrating key features, team members begin to confirm the project's viability in a concrete fashion. The first cycle can be thought of as a *proof of concept*.

Step 5: Assign Primary Components to Cycles

Once the overall objective of each of the cycles has been determined, core team members will assign components to the most appropriate cycle. A component should be assigned to the cycle in which the bulk of its development is to be accomplished, as was shown in Fig. 4.3. In traditional approaches to development, tasks are defined, then monitored for completion. Tasks are definitive and completion is a binary characteristic—completed or not completed. However, the real world doesn't work this way. A data model, for example, actually evolves over a significant period of time. The team may produce a preliminary rough draft, followed by a detailed model. Then, changes, modifications, and extensions will occur throughout the remainder of the project. Assigning components to cycles provides a way to monitor progress, but it is not a cleanly delineated process.

Particularly complex components may be completed during several cycles, although developers should take care to prevent these large features from being built too late in the project for effective correction to occur. In some cases, preliminary scope work may be done on a feature prior to its assigned cycle. For example, looking back at Fig. 4.3, we see that the Reorder Calculation component is assigned to cycle 3, but preliminary menu-selection items and possibly a rough screen layout of key data might be included in cycle 1 to demonstrate the scope of the application to the customer. For the size of our hypothetical project, there is no need to indicate this amount of work in cycle 1.

For cycle-planning purposes, the size of the component (or a specific feature that is part of a component) should be such that it can be implemented in two to three weeks. For cycles that are more than six months away, the size of the component chunk can be larger (with three to four weeks allocated for implementation). Suggestions for assigning components to cycles include

- Assign components to deliver usefulness to the client in every cycle.

- Assign to each cycle components that will help achieve the cycle objective.

- Assign components that implement major business operations functions first and decision-support functions later.

- Assign components to deliver feature breadth in earlier cycles and depth in later ones.

- Assign components to cycles based on their natural interdependencies.

- Assign components that are more risky and uncertain in earlier cycles.

- Assign components to balance resource utilization across cycles.

- Assign lower priority components to later cycles.

A cycle plan contains all of the components of the product, whether the component is executable code or user documentation. Figure 4.3 showed how different dimensions of the product might be addressed if the application were developed in four cycles. Figure 4.4 shows how two components, including "Reorder Calculation," might evolve over several cycles even though they are assigned to one. For larger projects, Chapter 9 demonstrates how to plan for additional component detail.

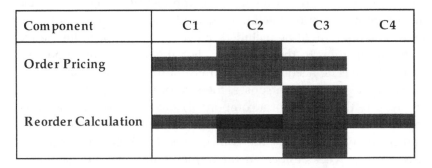

Component	C1	C2	C3	C4
Order Pricing				
Reorder Calculation				

Figure 4.4: Component Development Over Several Cycles.

Step 6: Assign Technology and Support Components to Cycles

A software development project entails much more than executable modules. Besides the code-related components of the product, a project also must produce documentation and install required technology components. Both are costly and time-consuming. Capers Jones (Jones92) has estimated 20 percent to 30 percent (the percentage is related to product size and type) of development cost is paperwork-related. While adaptive methods minimize formality of documentation to reduce work load, some documentation is part of every project.

The technology and support component sections of Fig. 4.3 showed how these items are assigned to cycles in the same manner as primary components. The suggestions for component assignment listed in Step 5 apply in this step also. Again, it should be noted that support components, such as a detailed requirements specification document, evolve over several cycles although the major portion of their development is assigned to a particular cycle.

Step 7: Develop a Project Task List

This step involves two options. The first alternative is to eliminate the step completely. One of the tenets of adaptive management is to define the desired results, the "what," and let the development team figure out the details of delivering the results, the "how." Progress is monitored at the end of a cycle or by reviewing large chunks of work (components completed) during a cycle. In a complex project, trying to monitor and control small chunks of work is counterproductive.

Managers experienced in more traditional project management techniques will feel uncomfortable with this "loose" approach. They want to be more in control, but as we have seen, being in control is a chimera. In fact, by having pieces of the application delivered every cycle rather than at the end of the project, the team actually has a better measure of progress than with more detailed traditional methods.

One reason the adaptive approach seems so different is that in traditional sequential life cycles, analysis or data design activities are performed during the corresponding life-cycle phases. In an adaptive approach, these activities will be performed during several cycles. There may, for example, be more analysis effort in early cycles, but analysis will occur in all cycles, as Fig. 4.5 shows.

The second alternative for this step is to begin with the task list created from the components and add additional detail-level tasks. For example, for the Inventory Picking component, tasks might be added to design, code, and test the component. There may also be occasional activities that are required for the project, but that are not directly tied to a component. Developing a project task list is often necessary in the transition to an adaptive approach. It provides some comfort to project managers and team members who have concerns about a less structured approach.

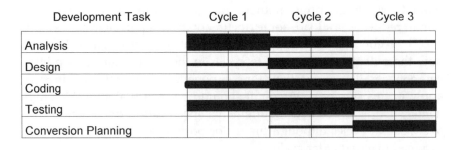

Development Task	Cycle 1	Cycle 2	Cycle 3
Analysis			
Design			
Coding			
Testing			
Conversion Planning			

Figure 4.5: How Development Tasks Span Several Cycles.

Similar to Adaptive Software Development in general, these seven steps should be viewed as iterative and open-ended rather than as a step-by-step recipe. The best strategy is to loop through the steps several times in the planning process. In numerous planning sessions with clients, my experience has been that, in the process of assigning components to cycles, the team gains a much better understanding of both the components and the overall project mission. The assignment process raises issues of project scope, component definition, component priorities and dependencies, size estimates, resource needs and availability, and more. Team involvement in the beginning speeds progress during the project.

Cycle Reviews

Since the environment of an adaptive project is turbulent and uncertain, the end of each development cycle must provide for reflection, status determination, and learning. While Chapter 6 will cover the Adaptive Development Life Cycle's Learn stage and learning techniques in detail, we should not leave the topic of adaptive planning activities without an additional brief discussion of the learning loop shown early in this chapter in Fig. 4.1.

The learning loop is specifically named to differentiate it from a feedback loop. Feedback connotes the mechanistic view of a closed-loop control system. For example, if we are controlling the heat source to a thermostat system, the only feedback we need would loop between the thermostat temperature gauge and some actuator to control the heat source. Managers are always looking for the simplest set of measures to "control" the development process. However, if we, as managers or developers, understand that our plans are speculative, then the understanding must also include the fact that simple measures are not enough to learn all we need to know at the end of a cycle. The information required to answer the question, "Are we tracking with the plan?" is much different than that required to answer the question, "Is our plan still tracking in the right direction?" An adaptive project must answer both questions.

Reviews at the end of each cycle should answer the following questions:

- Is the project on track if viewed from a broad scope, schedule, defect-level, and resource perspective? (In other words, What is the probability the product will meet its product mission profile?)

- Are the project mission artifacts still valid?

- Does the quality of each delivered component meet customer and technical specifications and expectations?

- Is the project team working efficiently and effectively?

In traditional projects, tracking schedules, tasks, and resources usually takes precedence over tracking components and evaluating their quality characteristics. In adaptive projects, tracking of components and their quality takes precedence. Built on the principle that *the application is the only acceptable model to the customer,* adaptive development projects track the delivery of actual software features rather than documents.

Much of the feedback gathered in traditional project monitoring originates within the boundaries of the project. The questions elicit information about how the project is progressing when compared to plan. Asking the question, "Are the project mission artifacts still valid?" requires entirely different sources of information. The project team must have mechanisms in place to scan the customer's environment and, in the case of software products for sale, the competitive environment. In the heat of trying to meet project schedules, people find it difficult to take the time to investigate these questions, but in a rapidly changing market, the investigation is essential.

A primary tool for reviewing results is the customer focus group. During focus-group meetings, customer decision-makers explore a working application in a facilitated environment. The deliverable results from the meeting are documented change requests.

Running focus groups well is one of the factors critical to achieving the desired results in an adaptive project. Focus groups perform the dual role of assuring product quality from the customer's perspective and catalyzing the organizational changes required to implement adaptive development. In a commercial software company, it is important to expose the product to review outside the development group. Whether the target audience consists of *actual* customers or *surrogate* customers from an internal department such as marketing, the exposure is critical to effective learning.

Good customers lead to good products. Too many internal systems projects fail because customers aren't committed. They want the result, but they aren't involved. Technical staff members assume customers don't become involved because they don't really care. The truth is closer to their not knowing. In principle, customers want to be committed and to be involved, but the processes, such as rigid, serial life cycles, established by developers actually get in the way. One of the key benefits of adaptive life cycles and customer focus groups is that they assist customers and developers in the "how" of partnership.

For many developers, and especially for those who originally viewed RAD methods as no methods, concern often exists about technical quality. There are several matters at play here. First, there is certainly a need for developers to maintain technical quality, but this is a concern regardless of the development approach used. Engineering is itself the process of managing trade-offs, and software engineering is no different. Technical inspections are a useful tool to insure quality, but developers must insure that *quality* isn't used as a mask for their own aesthetic sense of perfection. Second, important technical quality characteristics should be defined during project initiation and refined during the development process. Vague definitions of terms (such as "maintainability") are often used to justify elaborate and overly detailed design documentation with questionable linkage to the desired characteristic. If it is important, define it and include the characteristic as part of the project's specifications. If it is not defined, don't use it as an excuse.

While technical perfection isn't a reasonable goal, technical quality (as defined by appropriate specified characteristics) certainly is a goal. Technical quality is maintained, in part, through the use of technical inspections (reviews, inspections, and walkthroughs are all similar activities). During the cycle development process, inspections should be used when appropriate, instead of when the cycle is completed.

The third project management task that should occur at the end of each cycle is a review of team, process, and practice effectiveness. This review should be in the form of a project mini-postmortem conducted by the core project team. Questions should include: How well are the practices and processes working? Can we improve them in any way? Do we need to change any completely? Are the core team members working effectively together? These reviews will provide a mechanism for mid-course correction, and should prevent end-course panic.

Cycle Replanning

Because there is little chance the original plan will continue to be accurate, the cycle plan should be reviewed at the end of each cycle. Since adaptive projects are geared toward change, plans for them will require more revisions than those developed for traditional life-cycle projects. The only things that don't change are the cycle dates, unless their change is given team and executive sponsor approval. One of the objectives of the first cycle is to confirm that the original project size estimate was reasonably accurate. If the original scope and size estimates prove to be in error, the team needs to reevaluate the project and discuss it with the executive sponsor.

A Hypothetical Cycle Example

This section describes the kinds of components that might be allocated to a hypothetical four-cycle plan and provides some additional rationale for these allocation decisions. The project scenario to illustrate cycle planning was outlined earlier in this chapter, but important details to keep in mind are that it included a completed feasibility study, and that it is for a small to medium-size project (less than 1,500 function points), which has identified implementation and development technologies and an identified executive sponsor. Additional assumed characteristics are that the project is of moderate technical risk, that there are no extreme performance requirements, and that it has a typical internal client base (that is, one that does not have a software product for sale).

Describing a sample adaptive project plan involves a dilemma. A concrete example can enhance the learning process, but the problem is that a typical project can be misinterpreted. The unwary may use this sample as a model for projects in which another model would work better. For example, a project with a different set of risk characteristics would have a differently organized cycle plan.

The example plan contains four development cycles, called Cycle 1, 2, 3, and 4. The project initiation activity, assumed complete for this example, should have taken a week or two. Experience has shown that for larger projects with longer project initiation periods, it is useful to include a Cycle 0 for these project initiation and preparation (for example, training) activities.

The objective of describing the four cycles is to illustrate some of the thought process that goes into assigning components to cycles. The example is not complete, since characteristics such as component dependencies and resource balancing have not been considered.

Cycle 1: Demonstrate Project Viability

The objective of Cycle 1 is to demonstrate the ongoing viability of the project. The components developed in Cycle 1 provide a proof of concept. No matter how much planning the team does, projects are explorations into the unknown. The feasibility study, usually completed as part of project initiation, expresses the team's hypothesis that "Based on the information gathered to date, we believe this project is feasible within the objectives and constraints identified." Obviously, the proof of the hypothesis is a function of the final product. However, the objective of Cycle 1 should be to gather the right kinds of information to help prove the project is viable—that is, information that helps confirm the feasibility study's hypothesis.

In a typical project, two types of risk usually exist. First, the scope of the project, although documented in the project initiation documents, may not be well understood by either customers or developers; or perhaps the understanding of these two stakeholders is divergent. Customers may have a vague notion of what they want, or they may have a good *idea* about what they want but can't articulate it to the technical staff. Second, having had poor experience with results on previous software projects, they are often reluctant to commit themselves to the true collaborative effort required. The first risk is poor understanding of the scope; the second risk is poor collaboration. Because of poor collaboration, there is little chance for stakeholders to gain a better understanding of the scope. Addressing these two risks is critical to demonstrating the viability of any project.

Primary Components

Primary components compose one or more business features. These components deliver something useful to the customer, or at least provide background capability that allows another component to deliver something useful to the customer. Based on the identified risks, one objective of Cycle 1 is to verify the scope and concept of the product by the development of key primary components. In the project initiation activity, the project team gathered existing information about the project, developed an initial project objective statement, and outlined the purpose and features of the application. In Cycle 1, the development team must strive to clarify the project's objectives and scope by converting the preliminary feature-set requirements into a rudimentary first draft of the product and its appropriate supporting documentation.

The product of Cycle 1 is never pretty. It is usually hurried, buggy, and partially completed. The development team will want to delay the cycle completion until a higher-quality product can be produced; resist the temptation. When these components are delivered, the team will be able to confirm with the customer organization and the executive sponsor whether the project is viable and on track with the original scope expectations.

A second objective of this cycle is to engage the customer. The success of the project depends on the team's building a collaborative, open, communicative relationship with enthusiastic customers, but customers need to see concrete results with their own eyes. Customers will have been told about this new adaptive approach, but until the first focus-group review, they really won't have seen results. The public-relations effects of the early development and focus groups should be capitalized upon during the first cycle, in order to begin a cultural transformation.

During the development cycle, customer team members have definite tasks to accomplish. They are heavily involved in developing the specifications and in reviewing interim builds of the product. They also must concentrate on developing business scenarios by which they can demonstrate the features and test the system.

Development efforts for Cycle 1 should concentrate first on delivering the feature set specified for the cycle, and second on establishing a technical base for future development. With a first-cycle emphasis on scope and content, the development efforts might include

- menus and navigation screens

- screens for maintenance of key data entities

- management or operational inquiry screens

- key system integration components (such as, access to another application's database)

It is important to get as much visibly done as possible in Cycle 1, so initial work on a wide range of components is begun. The team must be careful not to become bogged down in technical areas. For example, if a difficult algorithm or database design issue leads to much discussion and consternation, look for some simple way to get the basic component done and put the unresolved problem on an issue list for the next cycle.

Technology Components

Technology components are the platforms on which feature components operate and by which developers ply their trade. In today's client/server, networked, Internet, data warehouse, e-commerce environment, the technology components can be extremely complex as well as extremely demanding on the project team.

Some of the timing constraints on the technology component can be daunting. Development begins the week after the initial kick-off and JAD session. There aren't months of analysis time during which to order and install, and then to get trained on, new technology components. The first features will need to be developed and then demonstrated within a few weeks. So, some technology components will need to be installed in the same cycle, often nearly concurrently with the primary component development effort.

There are considerations to take into account when assigning technology components to Cycle 1:

- What components need to be in place for development to begin?

- What components need to be implemented in order to operate the primary components scheduled for the cycle?

- What components must be implemented in Cycle 1 because work slated for later cycles depends upon their completion?

In less chaotic times, the implementation of technology components was often delayed until late in the project. Today, however, technology components need to be planned and implemented concurrently in order to keep pace with our multi-vendor, multidimensional environment.

Support Components

In most cases, Cycle 1 will not involve much work on delivery or installation tasks. There may be some initial planning required for these tasks, as in many IT applications where a significant conversion effort is anticipated. If so, a tentative conversion-issues list might be started in Cycle 1. If the final installation is to be across a wide customer base, some initial ideas about installation planning might be generated during Cycle 1.

Even with rapid development, there is such a thing as *too fast*. Cycle 1 should be used to establish some development guidelines and a foundation for subsequent development. With most high-powered development tools, spaghetti code or spaghetti data can be developed much more quickly than either could be developed using more traditional tools or languages. While a primary criteria for these tools is that they reduce the effort required to modify the application, initial setup can make a big difference in how effectively the tools operate. So, even in Cycle 1, the team needs to establish development guidelines such as naming standards, reuse approach, and data design considerations. Hopefully, the team is experienced with the tools and these guidelines are already in place and only need to be tailored to fit the project.

Cycle 2: Explore the Features

In the initial project plan, the team "speculates" on what the project's results should be. During Cycle 1, the team then builds components to confirm that the speculations were not too far off base. One of the objectives of Cycle 2 is to develop components in more depth—in essence, to explore the application. Cycle 2 (and often Cycle 3) is used to develop features that will make up the components, data related to the components, and the user interface all in greater depth than was done in Cycle 1.

Primary Components

Cycle 2 incorporates a more detailed exploration of components. A menu item and basic screen for a feature may have been developed in Cycle 1, but the feature will be specified in more detail in Cycle 2. Depending on the situation, the specifications can be gathered through additional mini-JAD sessions, through a more traditional interviewing process, or by a combination of discussion and prototyping with the customers who are members of the core team.

Changes will be made to features as a result of the first customer focus-group session. For many IT projects, it would be normal at this juncture to put additional work into the data modeling activity. Often, the focus group from Cycle 1 identifies complex or difficult areas in the data model that significantly differ from the original model. After the clients have seen the initial cycle results, they may have more enthusiasm for assisting with analysis. Often, functionality involving complex calculations or algorithms that were postponed in Cycle 1 are undertaken in Cycle 2.

Development efforts for Cycle 2 begin to add both functional and technical depth to the product. Components assigned to Cycle 2 are developed and approved change requests from the previous cycle are incorporated into the product. Additional efforts are directed toward defining the user interface and removing defects. Testing activities are more rigorous, so, by the end of Cycle 2, the product will be stable enough for clients themselves to act as product demonstrators for the next focus group.

Technology components are treated in Cycle 2 much as they were in Cycle 1, and specific suggestions listed for Cycle 1 can be applicable to Cycles 2, 3, and 4.

Support Components

At this stage, developers will know a great deal more than before about design issues as a result of building the first cycle. The project team now should be willing to abandon much of the prior coding if new design alternatives provide significant quality enhancements.

Although features will change, key design issues should be decided by the end of Cycle 2.

During Cycle 2, the project team should begin to think seriously about delivery and installation issues. In addition, test plans, test cases, and test data should be developed and implemented.

Cycle 3: Refine Features and Insure Performance

The three main objectives of Cycle 3 are to refine components until they meet all client requirements, to insure compliance with performance specifications (such as response time), and to insure that defect levels are acceptable. As in other cycles, developing additional components and revising previously developed components are part of the work effort. Refinements often involve implementing features such as error-detection and correction.

Particularly in business applications, project teams tend to bog down in error-handling routines and miss the "mainline" flow. The mainline is defined as the execution of a series of business processes as they are intended to be performed if Murphy were asleep (that is, as if no errors were encountered during processing). During Cycles 1 and 2, developers should have concentrated on getting the mainline correct. In Cycle 3, project team members should concentrate on how the application might fail, from either a business or a technical perspective.

By Cycle 3, any reuse strategy put in place by the developers and any architectural components should be completely implemented; any common modules, rules, or database procedures should be in place; and most systems-level design should be finished.

During Cycle 3, project team members should make final their thoughts and plans on delivery and installation issues. Conversion programs, installation routines, training programs, final quality-assurance plans—all should be at an advanced stage.

Cycle 4: Finalize All Product Components

The objective of Cycle 4 is to produce a total product that includes all components. Some of these components (for example, user documentation) may have been planned or even started during a prior cycle, but final development and testing should be completed in Cycle 4.

A wide range of activities will be performed during this cycle, but precisely which ones depends on the type of product being delivered. For example, a commercial software product will require that extensive effort be devoted to development of such items as user documentation, to possible geographic localization, and to refinement of installation procedures. In-house software products may need work done on installation, training, and conversion, while small applications will require only minimal effort in these areas. Some of the details that might be developed or made final during Cycle 4 include

- limited new component additions

- change requests from previous cycles

- product components
    - help screens
    - on-line training material
    - operations setup
    - security setup

- service components
    - training
    - help desk
    - reference material
    - user guides
    - operations guides

- installation and conversion
    - conversion software
    - installation software
    - documentation

Any of these components requiring extensive work should have been started during a previous cycle.

The Evolving World of Components

Components, objects, classes, containers, frameworks, templates, patterns—the list goes on. In the 1990's, application architectures progressed from client/server to Internet application servers, and the terminology now has expanded to describe different facets of what was once referred to as object technology. Some software development groups build components—modules of code, with one or more objects—which have standard, published interface specifications. A component developer might be tasked with creating an Enterprise JavaBean (EJB) component, for example. Application development groups find, select, customize, and assemble components into containers. Multiple EJBs might be combined into a container that operates in a specific environment (for example, CORBA).

From an adaptive cycle planning perspective, the word "component," as mentioned earlier in this chapter, is generic. Adaptive cycle planning is *result*-oriented, not task-oriented. Whether the result is an Enterprise JavaBean or a container holding six modified EJBs, the cycle delivery process is similar. That is, it is similar, but it is not the same. For example, in the Learn phase of an Adaptive Life Cycle, a customer focus-group review might be appropriate to review a container that implemented a business function. It would not be appropriate to include end users in a focus group to review most EJBs; however, a technical review with the users of the EJBs would be an appropriate *learning* activity.

While there are several evolving software development methodologies—split between those for developing components and others for assembling components—the Adaptive Development Life Cycle is appropriate for managing either type of effort.

Summary

> Redpointing in climbing is a technique for learning by iteration, by making mistakes, correcting, and converging on an ultimate solution.

> The Adaptive Life Cycle is a way for developers to redpoint software projects by falling off, correcting, and ultimately producing a quality result.

> Cycle planning provides additional depth to the project's mission. By concentrating on features and other components, the team learns a great deal about the project, including information rarely gained from a more traditional task-oriented approach.

> The waterfall model focuses on flow; the adaptive model focuses on cycling. One key to managing adaptive development is monitoring whether the results converge (cycle) or diverge (oscillate).

> The Adaptive Development Life Cycle consists of speculation—initiating the project and planning iterative cycles; collaboration—building components concurrently; and learning—reviewing for quality.

> *Versions* create deployable products. *Cycles* create verifiable feature sets to demonstrate to customers. *Builds* create incremental features visible to the developers.

> Project initiation establishes the foundation-level mission components for the project.

> Cycle planning is two-dimensional, first focused on demonstrable features, and second on tasks.

> Risk analysis drives cycle planning. High-risk areas are addressed early in a product's evolution.

> Cycle planning consists of determining the number and schedule of cycles, determining objectives for each cycle, and assigning features and other components to each cycle.

> A plan for a four-cycle project might be

  • Cycle 1: Demonstrate project viability (which is, essentially, a proof of concept for the project).

  • Cycle 2: Explore the features (that is, develop primary features in depth).

  • Cycle 3: Refine features and insure performance (implement requested changes and secondary features; focus on improving performance and quality).

  • Cycle 4: Finalize all product components (implement final features; develop all the services, documentation, and conversions, required to create a final deployable product).

> Each cycle ends with a review process, which examines the quality of the product from both a technical and a customer perspective, and evaluates the team's performance during the cycle.

CHAPTER 5
Great Groups and the Ability to Collaborate

The first chapter in this book described part of a July climb of Mount Jefferson in Oregon's Cascade mountains. There were two extremely critical decision-making junctures in this climb. First, the ice wall was expected to be a short, steep rock chute, but we were attempting an early-season ascent and we found ice. In coming up against the wall of ice, we learned of something new that called for serious reassessment. We had to ask ourselves: Is this the only route up? Do we have the right gear? Does one of us feel comfortable (or at least not completely freaked out) leading the way? There was pressure to not stop to make this assessment, because the time remaining in which to reach the summit was growing short, and we were in shade, surrounded by ice and cold. But the pressure to hurry on was tempered by our awareness of the risks we took in making the wrong decision. We decided to push on.

Two hours later, well past the ice wall, we faced our second reassessment. We were high on the mountain with the summit only eight-hundred feet higher, but we stopped again, this time to consider completely abandoning the climb. We had started at 4:30 A.M., it was now 2:00 P.M., and we faced at least two to three hours of climbing to reach the summit. We could see more ice than we had anticipated on

A woman seeking a divorce went to visit her attorney. The first question he asked her was, "Do you have grounds?"

She replied, "Yes, about two acres."

"Perhaps I'm not making myself clear," he said. "Do you have a grudge?"

"No, but we have a carport," she responded.

"Let me try again. Does your husband beat you up?" he asked impatiently.

"No, generally I get up before he does," she said.

At this point, the attorney decided to take a different tack. "Ma'am, are you sure you really want a divorce?"

"I don't want a divorce at all, but my husband does. He claims we have difficulty communicating."

—Source unknown.

the remainder of the route, and we knew that this could slow us down even more. We faced a four-to-six hour descent to high camp, an hour to pack up, and then a six-mile walk out to the car with full packs. Choosing whether to continue up or to descend was not an easy decision. No climber likes to abandon a summit attempt; however, in this case, the consequences of continuing on could be severe. If night fell while we were still on the mountain, we would be subject to frostbite or worse by spending the night out in the open with no sleeping bags or tent.

Climbers don't just climb. Technical climbing skills are just one part of the equation. Climbers must also have communication and joint decision-making skills. Since most climbing involves small groups (ours was three people), reaching the top is a process of continuous discussion, agonizing decisions, collaboration, and learning.

Michael Schrage (Schrage89) has described collaboration as an act of shared creation or discovery. Climbing, like software development, is also an act of shared accomplishment. The intensity of a climb and the realization that a wrong decision could have disastrous consequences leave little room for error or playing games. As the chapter-opening conversation in the sidebar indicates, communication in general can become frustrating and counterproductive very quickly— shared creation is even more difficult. Collaboration, or shared creation, requires more than listening; it requires negotiating sustainable decisions that each participant supports. In the pages that follow, we will look at ways the concepts and practices of collaboration can make the journey less burdensome.

If the intent of the collaborative effort in ASD is to create an environment in which groups will be able to produce emergent, innovative results, then our discussion might begin with questions about agents (people) in a complex adaptive system:

- What is the definition of a high-performance team?

- What kinds of people fit best?

- Is there a preferred leadership style?

- What are the shared values necessary for participatory interaction?

- What are the barriers to and motivators for collaboration?

- What is the best organizational structure for adaptive work?

- What are some practical collaboration techniques?

- What does complex adaptive systems theory contribute to the understanding of collaboration?

This chapter and the next five chapters investigate the interpersonal, cultural, and structural issues of collaboration. Interpersonal issues—those associated with how small groups interact effectively—are treated in this chapter. Chapter 6 addresses additional aspects of the Learn phase. Then, Part 3 begins with a discussion in Chapter 7 of management issues in the context of complex environments. Cultural issues—those associated with creating an environment supportive of emergence—are treated in the context of adaptive management in Chapter 8. Structural issues—those associated with creating and tuning an information infrastructure and building a network structure designed to foster collaboration across groups separated by time and space—are the topics of Chapters 9 and 10.

Before continuing on with the topic of collaboration, let me add a cautionary note about individuals and groups. Overall, this book focuses on interactions in groups, not on building individual skills. However, having individuals with high software-engineering skill levels is a critical success factor for adaptive groups. Drawing on our climbing analogy, we can see that teams whose members collaborate well, exchange ideas, share leadership, embrace diversity, and learn from mistakes still won't get to the top of K2, the second-highest mountain in the world and one that is more technically challenging than Mount Everest, unless each member of the team possesses world-class "technical" skills. So even though the focus of this book is on collaborative skills, we cannot lose track of the fact that technical software engineering skills are also critical to success.

Barriers to Collaboration

One of the questions asked in the previous section was, What are the barriers to and motivators for collaboration? One barrier is our legacy of Command–Control management style in which communication up and down the management hierarchy is much more important than cross-group, horizontal communication. A second barrier is our culture of individualism, which is not a barrier per se, but becomes one when carried to the extremes our culture often encourages.

"The western culture and western character with which it is easiest to identify exist largely in the West of make-believe, where they can be kept simple. . . . There is a discrepancy between the real conditions of the West, which even among outlaws enforced cooperation and group effort, and the folklore of the West, which celebrated the dissidence of dissent, the most outrageous independence."
—W. Stegner [1987], pp. 65, 69–70.

Wallace Stegner, a Pulitzer prizewinner and one of the foremost writers about the West, spent a lifetime trying to set straight the widespread romantic ideas about individualism in the West popularized by dime novels and shoot'em-up movies.

The Western myth is characterized by an attitude of "every man for himself" and by reliance on individual effort. The myths of software development are awash with counterparts to Billy the Kid, Bat Masterson, Wild Bill Hickok, and Annie Oakley. Bill Gates, Dan Bricklen, Gary Kildall, Mitch Kapor, and numerous others began the rugged, iconoclastic, individualistic, and antibureaucratic myths burdening today's software organizations, just as the myth of Western heroes burdened progress in the West.

We talk about collaboration, communication, and diversity, but much of our actual behavior is still individualistic. Our systems still reward individual rather than group performance.

If collaboration is an act of mutual creation, then valuing the contribution and participation of others is key to making collaboration a success. Unfortunately, organizations with the smartest people often have the most pervasive not-invented-here syndrome because they "flip the bozo bit" (McCarthy95) on other individuals and other groups. If I consider someone else to be incompetent (a bozo), then I have no incentive to collaborate with him or her. Furthermore, I won't trust such a person's work. By declaring others incompetent, I can also cover up my own poor collaboration skills.

In a recent workshop I held for a major software developer, the issue of collaboration was raised in an unexpected way. I asked attendees to list the most pressing problems currently faced by their product groups. Their problems included how to handle

- integration of components developed by other groups

- components received from other groups that don't work as expected

- dependencies on work done by other groups

- prioritization of activities when developing components upon which other groups depend

- vendor management

Later in the workshop I asked group members to identify and rate processes, scoring highest those they viewed as fun, and lowest those they viewed as tedious. On a scale of 1 to 100, each of the following processes was scored lower than 30:

- meetings

- presentations

- documentation preparation

- coordination of tasks with other groups

- coordination of component interdependencies

What this confirms for us is that most software developers today still prefer to perform assignments as individuals and regard communicating and collaborating with others as time-consuming and tedious. As the rating exercise showed, all the *fun* activities were those concerned with coding, testing, and other individual and intra-group activities.

In many organizations, everything depends on the skills of the individual. Just as it was in the Old Wild West myth, collective performance is sacrificed at the altar of individual excellence. These organizations don't think people have to collaborate, because it is assumed that each person is responsible for producing his or her own work and that group interactions will just slow progress.

Command–Control management styles are a second barrier to collaboration. Organizations that utilize a Command–Control style exude the characteristics of rules, predictability, and hierarchy, because they regard individuals as interchangeable pieces. In such organizations, there is not much need for collaboration, because managers prescribe what is to be done, and the "pieces" perform based on this. There is a need for communication, but not for collaboration.

"The problems of dealing with diehard independents who would rather quit than collaborate are major issues facing software development groups today."
—L. Constantine [1995], p. 47.

The Essence of Great Groups

"Great Groups" (Bennis97), "High-Performance Teams" (Katzenbach93), "Jelled Teams" (DeMarco99), and "Feature Teams" (McCarthy95) are all terms used to define teams that perform beyond the norm. However, the concept of teams has been overworked in recent years. As projects get larger and more complex, relationships both inside and outside our core teams determine whether a team is innovative and emergent or merely a group of people working together.

"None of us is as smart as all of us."
—W. Bennis and P. Biederman [1997], p. 1.

There is a tremendous emphasis in many organizations today on teams. This is often stated as a goal of moving from a hierarchical, top-down organization to an empowered, team-oriented organization. Team orientation is one of the foundation blocks for many process improvement efforts. Software development organizations seem to be good candidates for team orientation, but building teams takes enormous skill, discipline, and, yes, luck!

Given the proper environment, results can be startling. Teams can just as easily degenerate into being less than the sum of their parts. So what ingredients go into making successful teams? In Jon Katzenbach and Douglas Smith's *The Wisdom of Teams*, a "team" is defined as

> . . . *a small number of people with complementary skills who are committed to a common purpose, performance goals, and [a common] approach for which they hold themselves mutually accountable.* —J. Katzenbach and D. Smith [1993], p. 45.

"A jelled team is a group of people so strongly knit that the whole is greater than the sum of the parts. . . . Once a team begins to jell, the probability of success goes up dramatically. The team can become almost unstoppable, a juggernaut for success.
—T. DeMarco and T. Lister [1999], p. 123.

"[S]mall number of people with complementary skills. . ." The best teams are small; a core team should have fewer than ten members. The intense interaction necessary to enable a team to jell is difficult with larger groups. In addition to being a manageable size, every team must have the right blend of skills—technical skills, business skills, problem-solving and decision-making skills, and interpersonal skills. Selecting people to work together as an adaptive group is not just a matter of picking team members with *bandwidth* (intelligence); it is about selecting people with the ability to work collaboratively with other individuals and other groups. As product teams increase in size, and as teams consist of multiple feature teams, a different level of team building comes into play.

"[C]ommitted to a common purpose, performance goals. . ." Katzenbach and Smith maintain that their other recommended team-building actions will not produce the desired results without the right purpose and goals (mission) to which the team is committed. For a software product, the *goal* is related to the product's mission profile; the *purpose* should link the team to a deeper, more passionate commitment. While many teams do search for a philosophical purpose, Warren Bennis and Patricia Biederman offer their perspective, "Great Groups always have an enemy. . . . Whether the enemy occurs in nature or is manufactured, it serves the same purpose. It raises the stakes of the competition, it helps your group rally and define itself . . ." (Bennis97, pp. 207, 208).

Teams can squander the advantage of a common purpose and goals if members constantly argue over approach. If one faction believes in technical reviews and the other does not, or if one faction believes strongly in an adaptive approach and the other holds firm to a traditional approach, the team will not jell. In a jelled team, approach issues are discussed, decisions are made, and members support the decisions even when they may not be in complete agreement with the decision.

"[F]or which they hold themselves mutually accountable." In a society that stresses individuality, and particularly in organizational cultures that emphasize individual accountability, team members often find it very difficult to subordinate *their* performance to the team's.

Truly jelled teams develop their own sense of reward because they believe in their project's vision. Team members are willing to subordinate their personal agendas and hoped-for rewards to those of the team, and thereby provide the foundation that allows the team to outperform a group of individuals. It is not easy. Building the trust necessary to achieve mutual accountability involves hard work and the personal risk associated with resolving conflicts. Many groups avoid this risk and thereby deny themselves the chance to be a real team. Building high-performance teams begins with individuals willing to assume this personal responsibility.

The following paraphrase of Larry Constantine (Constantine95) identifies four types of teams and the environment in which they work best:

- Traditional hierarchical: The territory is familiar, sufficiently predictable, and the requirements are reasonably well-known.

- Breakthrough: The project must develop a major new innovation. Breakthrough project environments are chaotic, random, and iconoclastic, and the project structures can become very unstable as they get larger.

- Synchronized: The project's success is dependent on shared vision and common values. "The key to effectiveness . . . is full commitment by all members to a sufficiently complex and well-articulated vision of the mission and methods of the group" (Constantine95, p. 77). We often think of Japanese teams as well synchronized.

"Great Groups are not realistic places. They are exuberant, irrationally optimistic ones."
—W. Bennis and P. Biederman [1997], p. 16.

- Open: The territory is turbulent and fluid. Success requires an open-ended, flexible approach. "Adaptive collaboration is tailored for technical problem-solving. . . . What is important in this view of projects and progress is the adaptive fit between how the team is working and what it is they are working on" (op. cit., p. 72).

"Great colleagues bring out the best in each other. People ratchet up one another's standards and performance levels."
—M. Treacy and F. Wiersema [1995], p. 97.

While a breakthrough team structure would work for a small adaptive project, my view is that open teams are the most appropriate for adaptive projects in general—for two reasons. First, adaptive development requires the ability to scale up to larger, more complex projects—and open teams can do this. Breakthrough organizations work best with a small, collocated team. Second, a breakthrough organization is an excellent choice for creative inventions (in science, for example), but less suitable for the work of turning a creation into a complete product, such as in engineering. So, while there are exceptions to every rule, an open organization seems best for adaptive projects. Collegiality flourishes in open environments.

Technical climbing is partially defined by the connection of team members by a rope. For safety, in ice and rock climbing, one team member is anchored while another climbs. In glacier travel, the rope is protection against the climber falling into a hidden crevasse. In either case, roped-together team members protect each other against injury or worse. To "rope up" with someone expresses trust and confidence in his or her skills. Yet we often rope-up and commit our projects, our careers, and our organizations with little real thought about the other person's skills and even less thought about how the individuals might coalesce as a group.

In the squeeze of daily competitive pressures—higher quality, more features, less time—software development has increasingly become the realm of tools, techniques, methods, procedures, and processes. All good and necessary, but all doomed to partial success unless more emphasis is placed on the individual and on effective, committed, collaborative teams.

Using Complexity Concepts to Improve Collaboration

Margaret Wheatley, in *Leadership and the New Science* (Wheatley92), approaches the application of complexity science to management from a philosophical perspective. Wheatley's "new science" provides insight into the subject of participation and relationships in organizations. When organizations are viewed as *machines*, their management is more concerned with defining what the machines are to produce, and which cogs in the machines need to accomplish which tasks. When organizations are viewed as *complex living organisms*, their management views participation and relationships as vital to success.

Steeped in a mechanistic viewpoint, management has long been leery of humanistic techniques—from the transactional analysis of the 1970's to the move of the 1990's toward learning organizations. However, the current move to participative, human-centered management isn't just another fad; it is a fundamental shift, one supported by this new science of complex systems.

Wheatley's focus on participation and relationships gives rise to a different set of key management issues—a shift from what we do to how we facilitate:

> *How do we get people to work well together? How do we honor and benefit from diversity? How do we get teams working together quickly and efficiently? How do we resolve conflicts?*
> —M. Wheatley [1992], p. 144.

At the heart of complex adaptive systems theory's relevance to software development is the concept of emergence, and the factors leading to emergent results. In complex situations, project managers should become more focused on defining a mission and building relationships than on prescribing tasks.

Relationships run the gamut, from dysfunctional placating ones to hierarchical power-driven ones. But as the complexity increases in the environment in which a business or a software development team operates, certain kinds of relationships produce better results. An environment that encourages emergent outcomes has certain characteristics, and each of these characteristics provides its own management challenges.

"[T]he new physics cogently explains that there is no objective reality out there waiting to reveal its . . . 'reality.' There is only what we create through our engagement with others and with events."
—M. Wheatley [1992], p. 7.

Control Parameters

Ralph Stacey (Stacey96) defines five control parameters, which are characteristics of networks of people (teams, for example) that determine whether an organization is at the edge of chaos. These control parameters are the rate of information flow, the degree of diversity, the richness of connectivity, the level of contained anxiety, and the degree of power differentials. Emergent, or innovative, results are necessary to solve complex problems. Since complex adaptive systems concepts indicate that emergent results happen at the edge of chaos, we need to understand how groups bring themselves to the edge and what parameters they have to manage in order to create an environment more conducive to innovation and emergence. The following paragraphs explain these control parameters.

Rate of Information Flow. Rate of information flow means speed, not density. In fast-moving industries, the rate of information flow can overwhelm a software development team. Technology changes alone can rip projects apart time and again, as someone recommends, for example, use of a newer version of the Java development environment or use of the new application server that will be available in three months. Internal flow can be just as devastating. Changes in key customer requirements fail to flow through to the development team or they appear with mind-numbing frequency. Problems arise when the feature team in charge of the user interface fails to pass along the latest changes to the test team.

As complexity increases, managers in traditional organizations tend to restrict information flow by instituting rigorous processes to "control" the turbulence. In so doing, they overly restrict one of the crucial ingredients in producing emergent results. The adaptive organization, recognizing the importance of these flows, is better able to balance this temptation to be overly restrictive (this topic is covered in greater detail in Chapter 10).

One of the reasons that meetings in general, and collaboration in particular, have such a negative connotation for many people is that people fail to differentiate between two fundamental kinds of interactions—shared creation and information transfer. By confusing the two, groups fail to target techniques for better managing both types of interactions.

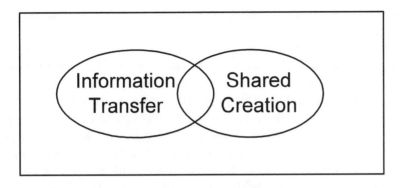

Figure 5.1: Information Usage.

As Fig. 5.1 shows, some overlap between the definitions does exist. However, information transfer is associated with presenting information—such as administrative or completed project information. Presentation information is not intended to be modified, updated, or manipulated by joint action. Its purpose is to inform. Shared creation adds value by producing a better product through active collaboration.

Michael Doyle and David Straus (Doyle76) suggest that there are several types of shared-creation and information-transfer meetings: problem-solving, decision-making, reacting and evaluating (feedback), and reporting and presentation (feed-forward). The first three are categories of shared-creation meetings. Meeting techniques need to be matched to the type of meeting in order to make each effective. For example, having a specific agenda and limited opportunity for discussion are very important to feedback meetings, but they are detrimental to problem-solving or brainstorming meetings. On the other hand, discussion of details during a technical review should be limited because the objective is to *identify* defects, not to *correct* them.

Most organizations abuse information transfer—they have too many meetings, meetings last too long, they waste time, and they sour people's views of all meetings, and therefore of collaboration, so much so that efforts at shared-creation interactions are shunned also. It is important to ensure that team members understand the difference between information transfer and shared creation so that the two processes can be managed differently. For example, whereas information necessary for shared creation is more effective if exchanged in face-to-face interaction, that information necessary for transfer is almost as effective on paper or in electronic form as it is in face-to-face exchange.

The ideas and practices set forth in this book are oriented toward improving information-sharing interactions, with emphasis on how to make meetings more effective, faster, and a lot more fun.

Degree of Diversity. Diversity brings spice to the creative process. It provides the breadth of creative ideas needed to produce complex products. Diversity takes many forms—technical skills and experience, cultural and racial backgrounds, personality types and temperaments, business skills and experience. Diverse ideas provide the richness needed to design highly competitive products. The wider the diversity, however, the more difficult the process of building a collaborative environment and producing results. Diversity brings new ideas, but too much creates differences of opinion, which make facilitating convergence more difficult.

Richness of Connectivity. Richness of connectivity can be seen in the number of connections between people or teams and the type of data flow. For example, graphic and audio clues convey a different type of information than is conveyed by plain text. Connecting two groups with a high-volume "pipeline" allows more data to be transferred, increasing the richness of each connection. Increasing the number of interconnected groups increases the diversity of the information exchanged in problem-solving. Too few connections produce stagnation; too many, instability.

"What the research indicates, then, is that at some critical point in the ranges of weakness to strength and many to few, the network is likely to produce great variety in behavior."
—R. Stacey [1996], p. 181.

One of the most dangerous actions in mountaineering is that of switching from leading a climb to belaying a partner below. In the congenial atmosphere of a sport-climbing area, the switch isn't usually a problem. But in the mountains, trying to communicate from a hundred feet away, around boulders and through cracks in the rock, can be daunting—especially when the wind destroys verbal communication.

At the changeover point, the lead climber stops, places anchors, and becomes the belayer. The climber at the bottom removes the lower anchors and begins to climb. Through shouted commands such as "on belay," "off belay," and "climbing," each climber knows what actions to take and the climbers advance safely. But how do climbers know what to do when the wind destroys communication? By understanding the context (that is, the circumstances surrounding the climb), and by substituting rope-tug signals (not always easy), and by listening to sounds even though the actual words are blurred, most climbers are able to safely negotiate the switch.

This climbing scenario illustrates that connectivity is about more than merely pushing data through a digital pipeline. Connectivity is a function of both content and context—a subject described more fully in Chapter 10. Content comes from the data while context comes from the information and experience that help the recipient interpret and understand the data.

Because the teams are dedicated and collocated in RAD projects, team members can establish high-context environments. Context is created by people sharing working papers, jointly creating and revising drawings, exchanging ideas in lunchtime chats and through hallway requests for explanations, and in team discussions. The team members know who produced the information, how complete it is, what team members can do and shouldn't do with the information, and who needs to look at the information before it gets distributed more widely. These contextual components are nearly invisible when team members are in close contact. As organizational and geographic distances are added, the problem is intensified. We must explicitly add context or the acts of shared creation are too difficult to sustain.

Level of Contained Anxiety. Working in turbulent, complex environments causes anxiety in most people. Our natural reaction to anxiety is to want to eliminate it. By containing anxiety, rather than trying to eliminate emotional distress, we use it to our advantage. We have all read of creative artists and scientists, from van Gogh to Einstein, whose eccentricities seemed to define their life and their work, and whose tension and anxiety may have contributed to their creativity and innovation. Complexity theory supports the view that anxiety is a cauldron for creativity, and I have seen in my own work that low levels of contained anxiety lead to stagnancy and inertia, and result in an organization's inability to adapt. High levels of anxiety lead to instability and lack of progress. Worse than either extreme is the situation in which projects oscillate back and forth. Innovative groups seem to be able to balance on the precipice, without becoming either comatose or psychotic.

Degree of Power Differentials. The introduction of participation and empowerment as management practices seems, at times at least, to eliminate power as a management tool. Ralph Stacey argues that the exercise of power should be a balancing act—too little exercise of power leads to confusion and endless consensus-building, while excessive use of power restricts collaboration. Stacey writes,

> *In the spectrum ranging from concentrated power exercised in an authoritarian manner to equally distributed power that is hardly exercised at all, a critical point is reached where one can find both containment of anxiety through clear hierarchical structures and directing forms of leadership, on the one hand, and the freedom to express opinions and risk subversive, creative activity without fear on the other.* —R. Stacey [1996], pp. 182–83.

As with the other control parameters, managing the degree of power differential requires a balancing act. Command–Control management style may be stifling, but no leadership at all is equally devastating.

The Management Challenge

One of the themes of this book is that collaboration enables organizations to generate emergent results. Furthermore, organizations are complex adaptive systems in which emergent results occur in the turbulent zone at the edge of chaos. If the edge of chaos is to be more than an interesting tag line, project teams need to understand the "levers" that push them to the creative edge. Stacey's five control parameters are these levers. They give us insight into how to achieve this position at the edge and show us why it is so difficult to remain poised there.

Relatively speaking, managing in a stable environment should be easier than managing in an unstable one—information flow is moderate, uncertainty is low, information is less complex, people are less anxious, and power differentials are high so that authoritarian dictates are not often challenged.

Some would claim that managing in a chaotic environment is even easier—there is no semblance of organization and results are truly accidental. Holding the middle ground between stability and chaos is difficult. But the middle ground is where extreme projects must be; there is no other choice.

"Leadership is the ability to create an environment in which everyone is empowered to contribute creatively to solving the problems."
—G. Weinberg [1994], p. 80.

Building Collaborative Groups

A collaborative group's role is to create an environment in which emergent outcomes occur. Many writers and thinkers on organizational effectiveness extol the benefits of cooperation, but why is it so difficult to achieve in practice? There are three main reasons. First is the myth of individualism discussed earlier. Second is the fundamental mental model of *survival of the fittest* that still lurks under the veneer of organizational civility. And third is the relative unimportance with which mastery of improved techniques, practices, and tools for group interaction are viewed in many organizations.

While groupware is a hot technology, its potential is far greater than its current utility. In fact, much of the groupware technology has diverted attention away from people's understanding of how to work together and instead has focused attention on how to use the technology. Although groupware products often are called collaboration tools, more often they are designed as communications tools, intended to inform, not to encourage sharing and creative acts.

As collaborators, we face two challenges. The first is to learn how we can identify interpersonal issues that will enable us to work more collaboratively in face-to-face situations. The second comes once we improve our ability to collaborate face-to-face: How do we turn that experience into the ability to collaborate over large geographic or organizational distances?

Neither challenge is easy to master. Collaboration, whether in a twosome or in a larger group, requires thought, energy, and occasionally even a facilitator or therapist. As Jim McCarthy writes, "It's difficult to overstate the value of coherent communal thinking, and it's easy to understate the difficulty of achieving coherent team thinking" (McCarthy95, p. 23).

Successful collaboration faces another problem. Much of the impetus for collaborative, participative groups has grown out of the organizational development (OD) community. Sometimes focused more on feeling good than on doing good, OD has a reputation for being genteel, mild-mannered, considerate, amiable, and touchy-feely. Collaboration in extreme environments is none of these things; it is characterized by behavior that is rough and tumble, loud, argumentative, full of fiery passion, and enlivened by flaring egos.

The Groan Zone

How team members move from being individual spokespeople to a unified, collaborative body can be difficult to understand. In his book on group decision-making, Sam Kaner calls the transition from the divergent zone of the individual to the convergent zone of the team member the "groan zone" (Kaner96). Explaining the differences between stages of divergent and convergent thinking, Kaner shows how at first, whether during a JAD meeting or during the initial phases of a project, people's ideas diverge. Even though each person wants to contribute to success and to getting the project going, each wants to voice an opinion. Everyone has a different perspective, a different experience, or a different context to bring to the project. Each person's thinking is divergent, bringing diversity to the process, but not much agreement.

Convergence occurs as the group's individual ideas are integrated into a whole solution. The process of integration does not entail compromise, in which everyone gives up something and no one is happy with the result, nor does it mean that everyone is in complete agreement. What convergence means is that everyone has participated and will support the final decision. Kaner points out that the goal is not merely agreement, but "sustainable agreement"—a unified position that doesn't fall apart hours after the decision meeting has concluded.

Kaner calls this period between divergence and convergence the groan zone because it is the time during which team members groan and complain. The groan-zone name captures what happens in most teams, and is so much more descriptive than the terms used by other team progression models (for example, "forming, storming, norming, performing"). The groan zone could also be used to describe the transition zone at the edge of chaos—it too is a turbulent zone where innovative, emergent results are generated. As the complexity and diversity of team membership increase, the groan zone may take increasingly more time. For example, in a three-day JAD session, it may take two-and-a-half days for individual differences to begin to converge.

The transition from diverging to groaning has one especially distinctive characteristic: It is the difference between forming one's own opinions and understanding others' perspectives. In the divergent zone, most group members voice their opinions to make sure their own ideas are heard by the group. Their behavior during this time could be considered to be a type of presentation, during which members are less concerned with understanding each other than with selling their own ideas. In the groan zone, however, an individual digs behind other people's ideas to try to uncover their reasons, assumptions, and mental models. Presenting ideas is relatively easy. Trying to articulate one's rationale for ideas is more difficult. Trying to understand other people's rationale for their ideas in order to reconcile them with one's own ideas is more difficult still—and is the essence of groaning. The progression from diverging and groaning to converging occurs in tandem with the transition from presentation to discussion to dialogue.

Groaning dominates adaptive development. Difficult problems and wrenching decisions cause teams to spend time in the groan zone because of the required interchange, sharing, and resolution of ideas, assumptions, and viewpoints. Unless convergence is forced periodically prior to the end of a project, project members will not test their convergence until it is too late to forge good collaborative decisions. Adaptive cycles force these intermediate results, which may or may not be a final product, but which are demonstrable. Cycles help team members both to see an intermediate product as well as to realize that their collective efforts really are converging. I noticed this recently during a six-month project I managed for a client: In virtually every week other than the week after a cycle milestone, the team felt it was groaning all the time. But each cycle's deliverables were achieved. The groaning soon became associated with progress toward the goal—not always enjoyable, but at least comprehensible. One benefit from all the early groaning is that later cycles experience convergence more quickly.

While groaning permeates adaptive settings, the typical scenario in a traditional project is that the team may not groan at all. Because the end date is so far away and the intermediate products usually consist only of documentation, the team believes it is on target. Everyone does his or her own assignments and little groaning occurs. At the end, when there is no time left to work through the issues, all the divergent thinking surfaces. The race to a ship-deadline is too late to begin the collaborative process.

Core Values

Whether group members are collocated or have been dispersed to distant parts of the globe, they need a set of core values upon which the success of any collaborative effort rests. The degree to which these values are embraced and demonstrated in day-to-day encounters directly impacts the degree of self-organization and, ultimately, of product success. These core values are mutual trust, mutual respect, mutual participation, and mutual commitment.

Mutual Trust. Trust within an organizational setting consists of three main components—honesty, safety, and dependability. Honesty means telling the truth. Safety means a feeling among colleagues that they will not be rejected or ridiculed when they voice their ideas and opinions. Dependability means knowing others will accomplish the tasks assigned to them.

One view of trust is that it is binary—either we trust someone or we do not. In practice, however, trust is much more complicated. Trust is not something static; there is a *trust bank account* in which reserves are built up and drawn down as needed. When trust is high, we can make mistakes, make comments we later regret, or show anger or irritation, and the trust account will enable others to overlook our human foibles. Trust levels are built up by our showing respect to others, doing what we commit ourselves to do, being honest and forthright in our dealings, and by not playing politics. When trust levels are low, minor events can trigger major confrontations and group members will fail to fully participate. As trust increases, cooperation and collaboration move from an emphasis on winning and losing to one of both parties winning.

Trust does not mean sacrificing accountability, however. Just because we trust someone doesn't mean he or she is unerring or infallible. If we don't, or feel we can't, trust other people or other groups to do their part, we can quickly become mired in duplicating work and making disparaging analyses of others' capabilities. Holding others accountable is not an act of mistrust; it is a way of guarding against miscommunication and flawed expectations.

Collaboration calls for civility, not group harmony. Operating in extreme environments at the edge of chaos—awash in anxiety, deadlines, long hours, distress, and irritation—is not conducive to harmony. Controversy, argument, conflict, confrontation, wrangling, even bickering and quarreling, when handled correctly, become grist for the creative mill. In the heat of the moment, conflict can move from constructive to destructive, from defaming products to defaming compatriots. But with proper attention, even these *subtractions* from trust reservoirs can be managed.

Mutual Respect. Respect means we value others for their unique contribution, even if it is outside our own sphere of expertise. Mutual regard is easy with people like us—the real test of a person and a project team's character is whether mutual regard extends to those who are different. Technical people, especially really smart ones, tend to readily "flip the bozo bit." In psychological circles, this might be called arrogance.

Geoffrey Moore's highly acclaimed books explore marketing strategy in high-tech businesses. In *Inside the Tornado*, Moore describes the conditions that have ignited hyper-growth in a particular market such as the Internet in the past few years. In his concluding chapter, Moore summarizes his philosophical advice:

> *This commitment to trust is mandatory if the leadership strategies of consensus management and decentralized operations are going to work. . . . The paradox of trust is that by intelligently relinquishing power, one gains it back many times over. Once you reach your personal limits, this is the only economy of scale that can help. And hyper-growth markets will push you to your personal limits faster than most other challenges in business.*
> —G. Moore [1995], pp. 237, 238.

Mutual Participation. The effectiveness of open, collaborative teams improves as more members freely participate. One person's ability to participate may be different from another's, but any person standing at the edge of the dance floor reduces the diversity upon which final results depend. By assuring *mutual*—but not *equal*—participation, we assure that although not everyone has the same capacity or experience, everyone does have an opportunity to participate in some key ways.

By valuing participation, we guarantee every member's entitlement to freedom of expression, no matter how outlandish a view may seem at first. Attaining a level of trust where seemingly outlandish ideas can be raised is key to enabling creativity and innovation. Freedom of expression leads to diversity of ideas, which leads to innovative solutions.

Mutual participation does not mean simply giving people an opportunity to express themselves—it is a genuine commitment on the part of all team members to try to understand what others are trying to convey. One person's persistence in trying to explain a point no one else understands can provide a breakthrough solution.

Mutual Commitment. Members of a high-performance collaborative group have both an individual and a collective commitment to achieving the purpose and goals of the project. They are committed to the project's goals and to working together, and they share the responsibility for making the project a success. There is an analogy in mountaineering: A climbing team usually has a leader and possibly even several different leaders for different activities. But if anyone abdicates responsibility, that action can lead to disaster. While one person may have the primary route-finding responsibility, most competent climbers constantly think about where they are as well as about the terrain ahead and behind. A good leader will check with the others periodically to validate direction.

A situation in which some climbers tend to abdicate responsibility is while climbing with a paid guide—the climbing equivalent of a consultant. In talking with guides over the years, I have been amazed to learn their conviction that many climbers they've led would not be able to find their way back to aid were the guide hurt. Teams whose members shrink from sharing responsibility with each other or who attempt to place all responsibility on a leader are far less effective than teams who freely accept responsibility.

On software projects, shared responsibility means that everyone *owns* the result. Katzenbach and Smith (Katzenbach93) call this "mutual accountability." It is built on personal responsibility.

Collaboration's Pitfalls

Despite the many advantages, there can be a potential downside to collaboration. Working together is not some moral or utopian imperative. Collaboration can lead to sameness and groupthink that can breed restrictive practices and place barriers to shut out external information. Collaboration can create solutions that reflect so many opinions that the final results are muddled. It can slow projects through endless meetings, discussions, and involvement of everyone in everything.

Successful collaboration depends on the ability of collaborators to understand the human dynamics of people working together. The major concern most people have in working together is the belief that their combined effort equals less than they can accomplish as individuals; that is, 1+1 equals 1.5 instead of 2, whereas the objective of collaboration is to have 1+1 equal 2.5 or even 3—that is, to engender emergence.

Because of their experience with poorly run meetings, many people do believe that group activities lead to mediocrity. Larry Constantine reported on research at M.I.T.'s Sloan School that focused on the drawbacks of group interactions, "particularly the so-called risky shift and the counter-tendency of groups to pull towards a mediocre mean" (Constantine95, p. 6). The researchers concluded that groups were inclined to arrive at more risky positions than individuals and that they produced results tending toward the averaging of ideas—compromise positions. However, Constantine points out that the negative outcomes reported by these researchers were in fact highly dependent on *how the group was led*. "Consensus design and decision making is at its best when the solution derives from the talents of all team members and reflects the experience, creativity, and critical thinking of all, not just an average of their contributions, but a genuine synthesis that combines their best" (op. cit., p. 7).

Successful projects exude commitment—to the mission, to the customer, to each other, to oneself. It is also difficult to succeed with a team that cares nothing about the outcome or to succeed with a project sponsor that blames development team members for problems, rather than helping them remove obstacles. And, it is difficult to succeed with a customer who wants no involvement other than receiving a finished product. Commitment is intensely personal. It is not a characteristic we can conjure up by performing magical, managerial, motivational tricks. The best commitment that people have to offer is self-motivated.

> **People *perform* work assignments; they *volunteer* commitment.**

People are not assigned mountains to climb—they willingly volunteer. Commitment to a project team arises from this same willingness in which participants find satisfaction and meet their own self-interests as well as those of the project.

To the general populace, mountaineers are often viewed as thrill-seekers. One question frequently asked of them is, "Why do you climb mountains?" The somewhat facetious but standard reply is, "Because they are there," but a more honest response is that they climb because it is an intensely personal and focusing activity. To do it for other than personal reasons would be folly. Few things focus your attention more than hanging off the side of a mountain, a thousand feet up. Getting to the top in mountaineering requires intense focus and commitment to overcome alternating waves of boredom, fear, and exhilaration, and a climber's success or failure is usually determined by his or her mental preparedness and determination to succeed. To get to the top in spite of the obstacles, fear, and fatigue that cause others to turn back is the climber's goal. Software projects are little different.

Rancorous Collaboration

If the success of a collaborative effort were completely dependent on mutually shared values and goals, its effect on product development would be limited. While shared values and goals act as a powerful force, and should be pursued avidly within a team, collaboration is an act of joint creation. It does not mean that team members must like each other, fully trust each other, or even have the same goals.

To understand how such seemingly negative factors can be overcome in collaborative efforts, let's look at the field of game theory, which delves into interactions between agents (people, organizations, countries) in their environments. One game, called the Prisoner's Dilemma, addresses interactions between divergent, self-interested agents and reveals the importance of an agent's knowing when to cooperate with others and when to compete with them. The original dilemma involved the decision alternatives available to two prisoners being interrogated by the police. There are two players in the game (two people, two groups, two companies, two nations) and four possible outcome combinations: Player A and player B may compete-compete, compete-cooperate, cooperate-compete, or cooperate-cooperate. Each player has the option of competing or cooperating, but neither knows what the other will choose.

The four outcomes reflect what happens in many real-world situations. For example, if one player competes and the other cooperates, the competing player wins big (is the cooperator a sucker?). If both compete, each gains a little. If they both cooperate, they both gain, but not as much as one would gain if he or she competed while the other cooperated. In some cases, mutual competition leads to mutual destruction as a third, presumably larger, competitor overwhelms both players. Cooperation leads to success against the larger competitor, but not to as much success as having the whole market to oneself. Sometimes, the seemingly best competitive strategy invites retaliation and, ultimately, results in less success.

Survival and success in competitive fitness landscapes require an understanding of which agents to compete with, and which ones to cooperate with. Sun Microsystems and IBM are competitors in many ways, but they may cooperate out of a fear that a third party will dominate important markets. There is enough mutual interest to override, at least in some areas, the individual's self-interest.

So, collaboration doesn't have to be genteel; it can be very effective as a rancorous interaction. It is not a necessity that collaborating groups share values and goals, just that they have enough areas of mutual interest to motivate a cooperative effort. Rancorous collaboration may not be the most powerful form of collaboration, but it is possibly the most prevalent.

Joint Application Development

Developing collaborative skills requires more than understanding the concepts; it depends on practices designed to help team members gain these skills in an organized way. Joint Application Development focuses on the concepts of collaboration; JAD sessions are collaboration tools that produce a variety of software development deliverables.

Originally developed at IBM in 1977 as an innovative way of developing installation plans for use by IT staff and clients, JAD has since evolved into a widely used technique. A typical JAD session might be described as

> **a structured, facilitated workshop that brings together client decision-makers and IT staff in order to produce high-quality deliverables in a short period of time.**

According to Capers Jones, JADs are a highly effective defect-prevention technique, providing consistently better results than are achieved by most traditional software-engineering defect-prevention techniques (Jones92). In addition to Jones's data, there are several good reference books on JAD, which explain everything from its benefits to the best way to arrange tables in the workshop setting. There are also good references for facilitation techniques, with Sam Kaner's *Facilitator's Guide to Participatory Decision-Making* being one of the best (Kaner96). My objective in this section, rather than to reiterate what is ably covered elsewhere, is to provide an overview of JAD practice in order to illustrate the concepts and ideas of collaboration and also to point out where I take exception to some common JAD "rules."

First, I offer modification of the above JAD session description:

> **a facilitated workshop that brings together cross-functional groups to build collaborative relationships capable of producing high-quality deliverables during the life of a project.**

While delivering a specific product is the intent of each individual during a JAD session, in many cases, the people in attendance do work together during other phases of a project. It is important in planning joint sessions, particularly in the early stages of a project, to give as much thought to building enduring, collaborative relationships as to producing specific deliverables.

What differentiates JAD sessions from a facilitated meeting? Nothing! JAD is a term used in software development (most commonly in IT organizations, less frequently among software vendors) but, in reality, it is a variety of facilitated meeting in which sessions may run longer than non-software-oriented facilitated meetings.

More specifically, JAD sessions are *feed-forward* meetings, intended to help participants extract information, solve problems, plan later work, and make decisions. They are useful throughout the entire development process. They generally are more open, more searching, and have more groaning in them than in *feedback* meetings, which are covered more fully in the next chapter.

For high-speed, adaptive projects, JAD sessions are required for early mission-setting and project-planning efforts, and for outlining requirements. There are too many people, with too many divergent opinions, to speedily arrive at a reasonable consensus on these items without JAD sessions. Sessions held for software vendors may be very different from those scheduled for in-house IT organizations with different objectives and different participants, but early collaboration is essential to speed. It is not how fast the project *starts*, but how fast it *finishes* that matters. False starts often lead to divergence, rethinking, replanning, and reworking, none of which contributes to finishing on time. A concentrated week at the start pays big dividends over the life of the project.

Facilitation

"Inspections without trained moderators will only have moderate success."
—T. Gilb [1988], p. 209.

Group meetings need a facilitator if participants are to get the most out of the time spent in a meeting. Sam Kaner summarizes the central role of a facilitator succinctly: "determining who talks when, and focusing the discussion" (Kaner96, p. 56).

Facilitators are specialists in group interaction. Project members may understand the need for a network guru, a Java specialist, a project manager, and others, but they often don't see the benefit in having a facilitator. Organizations that *do* use facilitators for meetings (such as

JAD sessions) should carefully consider expanding the role from meeting facilitation to collaboration facilitation.

The facilitator makes decisions about how to keep a meeting flowing. Good facilitators have bags of tricks, from icebreakers to decision-making tools, but a good facilitator always cedes to the group the right to question and change the meeting process. I once had a large group whose energy seemed to be lagging. When I suggested splitting into smaller groups to work on different issues for a while, the majority clearly thought it was a bad idea—we didn't do it.

Many managers and team leaders don't know how to use facilitators. Mired in Command–Control management styles, their egos are often too big to give up the least vestige of perceived power. But in open groups, power comes from respect, not from position. And, in fact, leaders who use facilitators know how much having someone else facilitate can enhance their own standing with the group.

By using a facilitator, the leader demonstrates that he or she does not always have to lead, but can cede leadership to others when appropriate. It is a fact that facilitating is very hard work; it doesn't leave much time to think about problems and decision criteria. A manager can often contribute more meaningful input as a member of the group, and not its leader, during meetings. Ceding meeting leadership to another is a subtle and effective way of demonstrating a real commitment to collaboration.

"[P]roject managers and official team leaders are probably the worst choice for leading any discussions or meetings directed at technical problem solving and decision making."
—L. Constantine [1995], p. 7.

JAD Roles

The four key roles in a JAD session are project manager, participants, facilitator, and scribe. While each role has certain characteristics and responsibilities, overlap and fuzziness of responsibilities can occur.

In general, the facilitator's role is to plan the session (in conjunction with the project leader), to orchestrate interactions during the meeting, to assist in preparing documentation, and to expedite follow-up after the session. Many published references on facilitation state rigid rules; for example, that facilitators are responsible only for the group process. The problem with this rule is that good facilitators often have a lot of *content* knowledge. Saying that they can't contribute content because it would compromise their *neutral* status eliminates an important source of ideas. Having facilitated hundreds of sessions myself, I always tell participants that if they want a content-free facilitator who deals only with group process, they should look elsewhere.

Another abused JAD rule is, "Leaders should never facilitate their own meeting." While agreeing with this in principle, and strongly advocating an independent facilitator, I don't think the rule is in any way universal. Over the length of a six- or twelve-month project, a facilitator may not always be available. While the leader has to be careful, facilitating a session can actually help with the longer-term goal of building collaborative relationships. When the leader takes on the role of facilitator, he or she relinquishes the role of leader. Since I, like many project leaders and facilitators, am prone to forget the assigned role, I give participants red flags to hold up when I slip from facilitating back into leading.

In general, the roles are well-defined. Participants are responsible for the contents of the agreed-upon deliverable. They have been brought together because of their experience and knowledge about the problem area. Participants need to have adequate decision-making authority over the contents of the specific deliverable from the particular session. Participant selection is critical to success. It is difficult for the wrong participants to produce the right product.

JAD sessions build both products and relationships. For the first of these, it is important to gather, organize, and distribute interim and final results—in essence, to document group memory. This is the primary job of the scribe and the secondary job of everyone else.

Roles are important, giving group members a focus for their efforts, but they can also be overly restrictive. Actors play parts. These are not who they *are*, but who they *play* for a while. Roles and role-swapping within a team enhance collaboration; it benefits the group for everyone to switch roles occasionally—even the leader and the facilitator. Roles can be used to construct and evaluate mission components, as well as to move from the general to the specific. Three generic roles team members can experiment with are visionary, realist, and critic.

Already a visionary (and true creative genius), Walt Disney was able to put himself in other roles during the evolution of an animated film. During the early stages of making *Snow White and the Seven Dwarfs*, he displayed his role-playing ability:

> *One evening in 1934 he gathered his artists together in an empty soundstage and, under a naked lightbulb, he acted out the entire story. . . . We are not talking here about a recitation. Walt apparently brought all the dramatic powers of an amateur Lon Chaney*

*to the drama. . . . The hours-long performance was the living
script the animators turned to again and again as they struggled
to complete the film.*
                    —W. Bennis and P. Biederman [1997], pp. 41, 42.

Disney was also a realist, a meticulous planner. He could dissect his
vision into specific, detailed components and establish plans to implement each. Disney was a critic as well, always asking what problems
might lie ahead and what risks were lurking. Roles can serve a group,
not as rigid constraints, but as identities that each person can try on for
a while. The critic can become a visionary, the visionary a practical
planner, the detail-oriented planner an idealist, the facilitator a participant. *Playing* roles can enhance the creativity of the group, build trust,
and increase the fun—what more could one ask?

Techniques for Successful JADs

Prepare

While there are myriad logistical details involved in preparing a successful JAD session (scheduling rooms in which to hold the meetings
possibly being the trickiest!), the most important parts of preparation
are deciding on the session objectives; researching background material; instructing the participants; and sleuthing for assumptions, mental
models, and political intrigue.

Having a solid session objective, and an agenda following from
that objective, is key to meeting the expectations for the deliverable.
Having a well-stated objective enables participants to prepare for the
session, allows both the leader and facilitator to plan the session strategy and length, and helps keep the discussion focused. While the session objective is usually jointly developed by the facilitator and the
leader, it should be validated with the wider group very early in the
session. A session objective might be something like, "Develop the
broad mission components for the project" or "Produce a prioritized
feature list, including brief descriptions of each of the identified features." One of the better uses of JAD sessions is to explore and define
the project's mission.

A second objective of every session should be relationship-building. Over the life of a project, building emergent environments is more
important than building any single deliverable. For example, a session
with an objective to define the product's quality profile is usually more

about uncovering everyone's individual mental models about quality than it is about specifying the particular characteristics. Uncovering, analyzing, and integrating each individual's view of quality will not be well served by a strict agenda. This is the time for an agenda item such as, "Allocate a half-day to discuss our philosophy of quality."

The facilitator is also a sleuth. Prior to a JAD session, particularly the first with a group or when new participants are invited, the facilitator needs to understand how certain individuals might affect the group. Several years ago, while planning a workshop for a software company, I mulled over how to approach a particularly vocal senior vice president scheduled to attend the session. I began my spiel, "Your involvement in this session is important as it shows concern at high levels of the company. . . ." He took advantage of a pause to interrupt me, "But you want me to shut up." After a joint chuckle, we resolved the issue. I said I wanted him to contribute, just not to dominate. He spent two full days with the group, listened well, made valuable contributions, and significantly improved his relationship with the developers.

Conduct the Session

How to conduct the session can be reduced to the basics mentioned under the discussion of facilitation—Kaner's "determining who talks when, and focusing the discussion." Much of what is important to conducting the session has already been covered, but additional concerns involve fairly run-of-the-mill practices for conducting any meeting—the logistical and facilitation details.

Produce the Documents

It is important to keep good records for the team. Both the final document and the intermediate discussion points are important to retain. Two months after a workshop, participants may find the intermediate points useful to understand not only what was decided, but why it was decided and what alternatives were considered. JAD sessions take considerable time and resources. Poor record-keeping diminishes the result.

Stable Change

Collaboration is the best tool for dealing with high-change environments. There is just too much for a single individual to manage. Collaboration also provides a stabilizing force.

Is the term "stable change" an oxymoron? Not really, for surviving and thriving in a hyper-growth market requires the ability to change, often and rapidly. Any company wanting to enter a hyper-growth market dealing with constant, unyielding, unrelenting change has to have a stable platform to which employees can retreat for solace and to re-stoke their engines. In *Managing at the Speed of Change* (Conner92), Daryl Conner defines "resilience" as the characteristic that enables individuals and companies to embrace change.

There are three keys to creating a stable platform, or "resilience" to use Conner's term, in a fast-moving software development environment:

1. *Establish respect and trust for leaders.* Walt Disney's animators had tremendous respect for his artistic sense. Walt couldn't draw the characters, but he knew when the results were right, and the staff respected and trusted his judgment. In many companies, the president might say, "We will now switch from going west to going north," and many employees at all levels of the hierarchy would merely sit and wait for the next pronouncement rather than take immediate action to move in the new direction. Trust and respect, much more than positional power, are characteristics of adaptable organizations.

2. *Build strong relationships.* Collaboration builds strong relationships. Especially in hyper-growth situations, relationships tend to be bonded in the heat of the fire. Those for whom the fire is too hot usually leave.

3. *Instill confidence in technical skills.* Technology changes, programming languages change, tools change, products change—but for the talented technical staff member, confidence (in oneself and in one's technical capabilities) carries one through crisis points and change. Stability lies in skills rather than in organizational position.

> *"[We] must refocus away from selling product and toward creating relationship."*
> —G. Moore [1991], p. ix.

Summary

➤ The act of collaboration is an act of shared creation and also of discovery.

➤ Collaboration is made more difficult in our culture because of the myth of individualism popularized by the Wild West of the 1800's cowboy and the 1990's silicon cowboy.

➢ Self-organization is the property of adaptive systems that enables a group to be more than the sum of its parts. Collaboration is the catalyst human organizations need to self-organize.

➢ "Great Groups," also called high-performance teams, have numerous identifiable, common properties, but one of the most difficult and illusive challenges to leadership is helping a group to jell.

➢ Larry Constantine identifies four basic types of teams, but an *open* type, based on collaboration, seems the best model for adaptive projects.

➢ Ralph Stacey identifies five characteristics common to organizations at the *edge of chaos* where emergent outcomes are most likely: the rate of information flow, the degree of diversity, the richness of connectivity, the level of contained anxiety, and the degree of power differentials.

➢ Collaborative groups, especially those in fast-paced environments, *groan* a lot. They struggle to create a product that converges on the mission profile. They struggle to integrate their own, and others', diverse perspectives.

➢ The core values to create win-win environments are mutual respect, mutual trust, mutual participation, and mutual commitment.

➢ Joint Application Development sessions are one of the most widely used collaboration tools in software development, particularly in IT groups.

➢ A JAD is a facilitated workshop bringing together cross-functional groups to build collaborative relationships capable of producing high-quality deliverables during the life of the project.

➢ Having good facilitators is important for effective meetings. Their interpersonal and meeting skills are usually in high demand in organizations understanding their worth.

➢ The four primary JAD roles and responsibilities are those of manager—responsible for group administration; participants—responsible for meeting content; facilitator—responsible for meeting process; and scribe—responsible for meeting documentation.

➢ As the pace of change quickens, the group needs to establish stable ground from which to confront the change. Overwhelming change, change without a stable platform, leads to chaos and psychosis.

CHAPTER 6
Learning: Models, Techniques, and Cycle Review Practices

Climbing, like most sports, requires practice, but it also requires variety in learning. Improving at any sport involves more than just going out and playing. Golfers who want to get better spend time learning on the driving range or on the putting green, competitive ice skaters spend countless sessions practicing figures, and basketball players spend hours drilling in basics like shooting free throws, for example.

Climbing requires much more than getting strong. While weight training can target specific muscles, it doesn't utilize those muscles in the same manner as actual climbing does. As with training in many other sports, training in climbing involves very specific techniques. Under the overall umbrella of strength are three specific kinds of strength, each requiring a different kind of training: endurance, power, and power-endurance. While all three kinds of training build strength, they also involve a movement-training component.

Endurance training is designed to keep muscles (most importantly, those in the forearm) working over a long period of time. Just like running long, slow distances trains the marathon runner or swimming laps builds endurance for the five-thousand meter swimmer, building endurance for climbing means spending long periods on easy terrain.

"The organization's ability to learn faster (and possibly better) than the competition becomes its most sustainable competitive advantage."
—A. De Geus [1997], p. 157.

Power training is geared toward climbs in which several moves require extreme muscle involvement to generate explosive force, like one repetition of maximum weight in weight lifting. The objective of power training is to involve a maximum number of muscle fibers. Working on a short, three-to-six-move climb in which each move is so hard that one never feels that the next move is even remotely possible is good power training.

Power-endurance training prepares one for climbs with many moves, none of which take maximum effort, but all of which are strenuous. Power-endurance training is the hardest (at least for me) because it is all about sustained pain. Power-endurance training is the equivalent of wind sprints for runners.

One of the three components of the Adaptive Development Life Cycle is learning. If we are to understand adaptive development, then we must learn about learning. One way to do this is by analogy with climbing, because climbing training illustrates two aspects of learning: First, it is not easy; and second, there are multiple kinds of learning.

What Is "Learning"?

The training techniques described above reinforce the concept that learning is often difficult and requires the individual climber to have persistence and dedication, the will to overcome boredom, the strength to face pain, and the maturity to resist the compelling desire to just go climb and have fun rather than suffer through the training. Members of an adaptive project feel similar emotions; rather than suffer through the difficulties of learning during each iterative development cycle, they would often prefer to isolate themselves and just code.

To some, climbing still may appear to be a sport solely about strength or endurance or strategy, but it is also about passion and emotion. It is about understanding oneself—one's goals, strengths, and weaknesses—and one's response to anxiety. A fearless climber is dangerous, because he or she probably doesn't know when to back off and try again another day. Climbing to—but not beyond—the limits of one's ability requires aggressiveness while overcoming anxiety and fear; it requires a strong mental discipline. Mental training is the essence of climbing.

There are two types of learning that are important to this discussion, each requiring a different kind of mental training and strength.

The first type of learning is called, by various authors, enhancement, maintenance, adhering, single-loop, technical/task, and training. The second is referred to as anticipative, double-loop, innovative, proactive, and transformational. I think of the first type of learning as *learning about things* whereas the second type involves *learning about oneself* (where "oneself" can be a person, a group, or even larger).

In climbing as in software development, there is always something new, a new wrinkle that pits existing ways against change. Some traditional climbers, for example, have difficulty understanding, or even tolerating, sport climbing (sport climbers use preset, protective anchors bolted into the rock, whereas traditional climbers never do). A traditional climber able to persist in the face of extreme physical discomfort becomes apoplectic at the prospect of anyone climbing in an *unapproved style*. The traditional climber sees "unapproved style" as endangering a carefully nurtured, learned model of climbing.

At the beginning of this section, I stated that learning is difficult and requires persistence and dedication. In the world of software, if a developer believes Java is absolutely the best programming language and also the wave of the future, acquiring information about Java requires persistence and practice but little anguish about the wisdom to pursue it. However, if another developer is a fervent C programmer and thinks Java is a short-term fad, there may be a significant *mental* barrier to learning. Mastering Java in the case of the first developer requires *training;* for the second developer, *learning* is required.

> *Training* **is the acquisition of skill or information.** *Learning* **is an attitude.**

The typical organization is still poor at learning about "itself"—its beliefs, assumptions, mindset, or mental models. Learning about oneself—whether personally, at a project-team level, or at an organizational level—is key to the ability to adapt to changing conditions. In *Microsoft Secrets* (Cusumano95), Michael Cusumano and Richard Selby identify Microsoft's ability to build a learning organization as one of the company's main strengths. They describe Microsoft's key strategy as continuous self-critiquing, feedback, and sharing. Stated slightly differently, Microsoft's practical approach to being a learning organization emphasizes, first, learning about operations, products, and customers, and then, broadly sharing information across the company.

Success is not about what we know; it is about how we learn.

In a traditional project, feedback is viewed as a deviation—something to be reduced either by improving the process, increasing the training, tightening the management controls, or by ignoring the data. In most instances, feedback necessitates doing some task over. Focusing on *redoing* means coming to terms with the realization that, no matter how hard we try, we can never get it right the first time and that adequate feedback is the only way to succeed. "Redoing" does not mean we have been sloppy, or that we don't care about making errors—good processes for reducing errors are still important. But (particularly as the uncertainty surrounding a project increases) redoing does mean that

Success is determined by the adequacy of the feedback, not by the accuracy of the feed-forward.

If getting good feedback is the focus of each cycle, project teams will put more energy and time into identifying where the feedback should come from, what types of feedback are valuable, and what mechanisms for using the feedback are needed. The ability to look critically at one's assumptions and beliefs—often beliefs that brought success in the past—is one of the most difficult undertakings of organizational learning. It means recognizing a Dilbert principle paraphrased in terms of oneself: "The greatest risk I face in software development projects is overestimating my own understanding."

It is not fun to face our own fallibility, but why should any of us think we can, as individuals, understand the complexity of an entire software project? To learn, we must be willing to make mistakes, to be open to others telling us about our mistakes, and to act on that information even when our feelings are hurt.

The phrase "even when our feelings are hurt" is key. No matter how much we might believe in the learning process, no matter how good our team is about working in a rich, collaborative environment, having our mistakes presented to us by others always takes a toll on our feelings and our egos. Strong teams enable us to work through those feelings to a positive outcome. Weak teams don't even take the chance—and consequently minimize learning.

Organizational learning is about relationships. If a developer is extremely competent in technical areas but has no meaningful relationship with the customer, the overall results will be disappointing. While a commitment to learning doesn't seem so hard, a commitment to building relationships is far harder, especially for many technical staff members. But a commitment to learning about yourself necessitates a commitment to relationships and to all the messy, irrational, complicated aspects that come with working with other people.

Organizations must put structures and techniques in place to encourage learning. "A learning journey" is a nice phrase, but how do we actually do it? Several specific techniques of adaptive development are oriented toward learning. For example, an Adaptive Development Life Cycle provides customer-oriented deliverables at regular, short-term intervals. Learning is enhanced by frequency and repetition.

Adaptive development encourages getting good feedback using learning techniques such as customer focus-group reviews, software inspections, and postmortems. Because there is little published material available on customer focus-group review techniques in comparison with widely available references on inspections and postmortems, I describe them in detail later in this chapter. First, however, are descriptions of two models, which further illustrate how learning organizations function.

Senge's Learning Model

While the practices that define a learning organization have been articulated by many authors, Peter Senge, in *The Fifth Discipline: The Art & Practice of the Learning Organization* (Senge90), vaulted the dialogue to near silver-bullet status. The book's opening paragraph emphasizes how crucial it is to see a system as a larger whole, rather than just for its pieces:

> *From a very early age, we are taught to break apart problems, to fragment the world. . . . but we pay a hidden, enormous price. We can no longer see the consequences of our actions; we lose our intrinsic sense of connection to a larger whole.*
> —P. Senge [1990], p. 3.

Senge details five approaches needed to create innovative learning organizations: systems thinking, team learning, mental models, shared vision, and personal mastery. What Senge and others, including Gerald

"Systems thinking is a discipline for seeing wholes. It is a framework for seeing interrelationships rather than things, for seeing patterns of change rather than static 'snapshots.' . . . Systems thinking is a discipline for seeing the 'structures' that underlie complex situations, and for discerning high from low leverage change."
—P. Senge [1990], pp. 68, 69.

Weinberg, have written on these topics are summarized in the paragraphs below.

Systems Thinking. In our fast-paced culture, we tend to react to events rather than seek an understanding of the system that has created these events. Weinberg's four-volume *Quality Software Management* series begins with a volume on systems thinking, which describes three fundamental abilities one needs in order to be a good manager of software projects (Weinberg92, pp. xiv–xv):

1. the ability to observe what's happening and to understand the significance of your observations

2. the ability to act congruently in difficult interpersonal situations, even though you may be confused, or angry, or so afraid you want to run away and hide

3. the ability to understand complex situations

The third of these, the ability to understand complex situations, essentially means taking time to understand the whole as well as how the pieces interact. "Systems thinking" entails thinking about the whole system and its ability to generate emergent results.

Team Learning. Teams are the vital building blocks for learning organizations. Small groups that find a way to create something greater than the sum of the parts, that sample the environment and use good judgment to make correct decisions, and that create a collaborative setting, are key to making the transition from individual learning to organizational learning. The concept of emergence is that innovation is a property of the interactions (the learning) among team members, rather than a property of some single contributor's action. Emergence begins with the team members' ability to learn from each other.

Mental Models. Mental models are formed from our understanding of and beliefs about how the world operates. Mental models shape our actions. In *The Fifth Discipline Fieldbook*, Senge and his coauthors define the concept:

> *Mental models are the images, assumptions, and stories which we carry in our minds of ourselves, other people, institutions, and every aspect of the world. . . . Human beings cannot navigate through the complex environments of our world without cognitive 'mental maps'; and all of these mental maps, by definition, are flawed in some way.* —P. Senge et al. [1994], p. 235.

Whether considering the relative merits of Monumental and Accidental development styles or those of results-based versus process-based techniques, our discussions reflect different mental models of the software world. People who believe strongly in one model, regardless of which one, often are oblivious either to the weaknesses of their own choice or to the strengths of another's. Mental models are at the core of learning and adaptation. Until people, or teams, or organizations are willing to confront their own mental model as potentially flawed—to question *themselves*—and to acknowledge that another model might offer improvement, the status quo will prevail.

By their nature, mental models can be rational, scientific, well-grounded, consistent, and methodical, or they can be illogical, inconsistent, unscientific, and haphazard. The only criterion on which individuals or organizations can base their mental models is a belief in the model itself.

Shared Vision. Senge's concept of shared vision is similar to the concepts previously discussed in Chapter 3, treating joint development of mission artifacts. A development team with a shared vision understands not only the words of the project objective statement, for example, but its members also understand the assumptions, beliefs, strengths, weaknesses, word definitions, and alternatives discussed in arriving at that project objective statement.

Personal Mastery. Mastery differentiates the beginner from the craftsman, the novice from the professional. "Personal mastery is the discipline of continually clarifying and deepening our personal vision, of focusing our energies, of developing patience, and of seeing reality objectively" (Senge90, p. 7). To me, personal mastery is the drive to learn more about one's profession of software engineering or software engineering management. Personal mastery involves taking the initiative for learning, whether it is through independent reading or through enrollment in a university seminar, rather than depending on an employer to provide all training.

Organizations that want to become more adaptive first must fully understand how individuals and organizations learn. Senge's model provides a good starting place.

A CAS Learning Model

Complex adaptive systems concepts provide a high-level framework for how organizations adapt to their environment. John Holland, in *Hidden Order: How Adaptation Builds Complexity* (Holland95), addresses

"Each software professional has a clear notion or 'mental model' of how to develop software. . . . Yet it is the limits of these models that undo them. It is not that one mental model is necessarily better than another, but that a mental model that works in one situation may not work in another. . . . Collectively, the diversity of mental models forms an organizational barrier to constructive change."
—B. Smith et al. [1993], p. 11.

learning within the context of any adaptive system, whether it is a natural biological system or an organizational one. Holland's work is more general than Senge's since it applies to any complex adaptive system, but it is a valuable reference because it provides additional insight into how organisms learn and adapt.

Whether the adaptive agent is a biological cell, an individual person on a development team, or a single business within a market, Holland proposes a general model that consists of a performance system, a credit-assignment method, and a rule-discovery process. The *performance system* is the set of rules that an adaptive agent uses to interact with the environment. *Credit assignment* is Holland's mechanism for keeping score; for example, by measuring profits or market share, businesses keep score in comparing themselves with competitors. *Rule discovery* is how an adaptive agent learns to compete more effectively.

Rules permit a response to the flow of information that the agent's environment produces. The following example gives stimulus-response rules describing how a frog might catch dinner:

> IF small flying object to left
>
> THEN turn head left 15 degrees, flick tongue.

In a business, a stimulus-response rule might involve the processing of an order form when a sale was completed. The interpretation of Holland's term "rule" should not be limited to algorithmic usage but should be broadly interpreted. These rules correspond to mental models in humans. In looking at the role of rules, Holland makes a very important distinction:

> *The usual view is that the rules amount to a set of facts about the agent's environment. Accordingly, all rules must be kept consistent with one another. . . . There is another way to consider the rules. They can [be] viewed as hypotheses that are undergoing testing and confirmation. On this view, the object is to provide contradictions rather than to avoid them.*
>
> —J. Holland [1995], p. 53.

Using the frog example, a contradiction could be phrased as

> IF small flying object to left
>
> THEN turn head left 15 degrees, flick tongue,
>
> EXCEPT if object is banded black and yellow and buzzes!

Whereas Senge alludes to the flaw in every mental model, Holland constructs a framework that indicates, first, that a "normal" or "default" mental model is relatively easy to establish (but it is not necessarily easy to convince someone of its relevance); second, that the exception conditions are more time-consuming to find and establish because they arise from experience; third, that contradictions exist all the time between normal and exception rules; and fourth, that learning (rule discovery) is a continuous process of sorting through and testing new sets of rules.

The science of complex adaptive systems provides two especially important insights into how development teams, or entire organizations, can be more effective at learning and adapting. The first deals with the magnitude of change undertaken; the second deals with the importance of exception conditions.

Innovation and Change

Companies attempt large-scale changes, such as Business Process Reengineering (BPR) projects, on a regular basis. Biological agents don't initiate extreme changes, but they attempt, and usually fail, to adapt when the environment changes drastically. As with the disappearance of the dinosaur, and the even more catastrophic late Permian event 225 million years ago when 96 percent of all living species vanished from the earth, extreme changes have a high probability of failure in biological systems (Gould89). Hundreds of species, many superbly honed as the fittest in their environment, died in the Permian extinction. Some, barely surviving in small ecological niches, accidentally happened to have characteristics allowing them to flourish in the subsequent Triassic period. The bigger the change, particularly in the external environment, the larger the chance of failure. Such survival could be called *survival of the luckiest.*

Adaptive organizations understand the danger in large-scale change (although sometimes there is no alternative). They also understand that innovative learning comes from allowing and examining inconsistent rules.

> **Innovative learning is about understanding the exception conditions to our general mental models.**

There are times, however, when our entire mental model is wrong, or at least woefully out of date. In more cases, the primary problem is that the general model is assumed to be valid under too wide a range of conditions. We often try to follow the usual rules, maintaining consistency, but understanding the role of exception conditions in learning leads to an insight about one of the software industry's favorite fantasies—silver bullets.

> **Silver bullets are new mental models for which the exception conditions are not yet understood.**

Mastery of a field of knowledge comes, in part at least, from knowing one's limits. Zealots do not believe in limits. Object-oriented programming zealots believe O-O is the one true way. Capability Maturity Model zealots believe the CMM encompasses all software development. Technical inspection zealots believe no program should leave home without one. Zealots don't recognize exception conditions, and are therefore doomed to run head-on into the fatal flaw contained in every mental model.

It is interesting that resistance to change usually surfaces first as exception conditions. My experience from years of implementing RAD projects is that many of the comments that people on a RAD project team will make relate to conditions under which RAD will not work. Participants concede that the *pilot* project worked, but claim, "RAD won't work for larger projects," or "It won't work for projects requiring integration with other systems," or "It won't work for maintenance projects." My response to these commentators is, "You are right. We don't have much experience with RAD under these other conditions yet. But under *this* set of conditions, it's a great addition to your development arsenal."

> **Big changes often lead to failure. Little changes often lead to big changes.**

The need for big changes comes about as a result of a disruption, either from a large-scale disruption in the environment (for example, a new technology) or from a failure to recognize the severity of a disruption that manifests itself over time. For example, mini-computer manufacturers failed to recognize the disruption caused by the advent of personal computers, trying for years to ignore PCs. When these companies finally were forced into attempting large-scale changes, they generally failed. There is no way to accurately predict the impact a large change will have on a complex adaptive system. The bigger the perturbation, the more likely the outcome will be unexpected.

The high failure rates of large Business Process Reengineering projects has been well documented. Started with grand visions, most of these projects failed to come close to their intended benefits. The big changes recommended by BPR projects attempt to overthrow broadly based, well-entrenched, mental models. Advocates for change start by declaring the old model wrong—thereby undermining people's sense of worth—and then substitute a new model and a new set of rules. The new models are often based on projections not experience, and therefore are subject to a high degree of skepticism about their validity, particularly when one's job is at stake.

What about approaching the argument from another direction? For any mental model, one can postulate a narrowly confined exception condition that others will at least consider. The statement, "Waterfall models work well, except for small, high-speed projects on which RAD works better" might be an exception condition that organizations would at least be willing to investigate, as opposed to the statement, "Waterfall models are useless; RAD is the new, better, and only way!" which engenders instant—and justified—resistance.

Once a group becomes comfortable with an exception rule for one set of conditions, the conditions can be extended. Perhaps RAD might next be suggested to work with small and medium-size projects. At some point in this evolution, even the most diehard Waterfall devotee will ask the question, "Is RAD the general case, and Waterfall the exception?" At some point, a seemingly insignificant new exception condition's acceptance may trigger this major transformation. Or, it may become evident that RAD really *is* an exception condition! Self-organizing systems, given the right environmental conditions, don't need to be forced into change—a gentle nudge is usually more effective.

One key to successfully making a large change is the nibble-nibble-bang approach—that is, to find those small changes with potentially large leverage. Since complex systems are nonlinear, we are never sure about the outcome, but building on small changes moving in the right direction can propel us into nonlinear, self-organizing, high-leverage adaptations.

An example of this nibble-nibble-bang approach to learning and change is Geoffrey Moore's high-tech marketing strategy (Moore95), in which he advocates picking a market niche and fully exploiting the niche with a total product solution before moving on to the next niche. At some point, when enough niches have been addressed, the cumulative effect begins to take over to create the "tornado" of change. Trying to create the tornado without first exploiting successive niches is a recipe for failure, according to Moore—but of course, there are exceptions.

Earlier in this chapter, the observation was made that learning is not the acquisition of skill or knowledge, but that it is also about attitude. Both Senge's learning model and the CAS learning models encourage new attitudes about learning and adapting in today's turbulent market environment.

Learning Techniques

Without specific techniques, learning models are not actionable. Often, techniques come first and real strategies come later. While everyone in business today touts the concept of being close to customers, few have solid techniques to make the concept a reality. The three techniques discussed next in this chapter are not new, but they address areas that every project should monitor and learn more about—customers, products, and team performance. These review-oriented techniques help implement the Learn component of the Adaptive Development Life Cycle's Speculate–Collaborate–Learn phases.

If it is true, as I stated earlier in this chapter, that the greatest risk we face in software development projects is overestimating our own understanding, then fast-moving project teams should be constantly anticipatory, asking themselves questions such as,

- What do we know?

- What do we need to know that we don't know?

- What do we not know we don't know?

- What do we know that is wrong?

Finally, we must ask,

- What do we *not* understand?

For answers to these questions, team members can apply any of three learning techniques, assessing their understanding in the context of additional questions the reviews help answer:

- customer focus-group reviews—What do customers think about the product being delivered? Does the product satisfy their business needs and other defined quality criteria?

- software inspections—What do the technical members of the team think about the product being delivered as measured by the criteria established in the quality profile?

- postmortems—What do team members think about their performance? How did practice compare with objectives?

A *customer focus-group (CFG) review* is a research tool, by which decision-makers explore a working application in a facilitated environment, resulting in documented change requests. Customer focus groups serve two main purposes: They enable development team members to gather information from customers about the interim product, and they assist developers in building better relationships with customers.

As described in the previous chapter, JAD sessions have been used extensively by IT organizations, but focus groups have not. Both my personal experience and studies reported in industry trade publications have shown JAD sessions to be very effective at helping project members gather information, make decisions, and build relationships. Customer focus groups cement these relationships by demonstrating results. JAD sessions offer the promise of results; focus groups deliver actual results. JAD sessions are used to gather application requirements in the form of documents or graphic models (data models, for example). Then, at the end of a cycle, a CFG review can be used to demonstrate a functioning application. JAD sessions and CFGs make a powerful duo.

The three cycle-review techniques presented in this chapter enhance collaboration: They are designed to facilitate cooperative acts of shared creation. Interpersonal values of trust, respect, and mutual participation apply, but which comes first—the practice or the trust? To answer this, I share two examples based on personal observation from which we can draw some conclusions. The first example occurred during a project several years ago when I saw a series of customer focus groups transform one of the crustiest, most negative, untrusting curmudgeons I've ever worked with into a still crusty, somewhat negative curmudgeon, but one who began participating in shared creation. The second example comes from my experience with a focus group scheduled at the end of the first development cycle in which a senior manager was ecstatic with the results generated in four weeks. The results far exceeded the manager's normal experience with any IT project. In both cases, the customers increased their trust in the IT project team; IT team members discovered that customers can be reasonable people; and both groups realized that the focus-group technique was an *enabler* to a better relationship. Techniques have to foster and support concepts—and vice versa.

In short, customer focus-group reviews ask, "Are we doing the right thing?" Software inspections ask, "Are we doing the thing right?" Postmortems ask, "Can we build the thing a better way?" When seen in the context of high-speed projects, these cycle-review techniques may seem to be too time-consuming. However, it is in part *because* of the speed that they are so important. The adaptive approach requires the team to go fast, to produce the best deliverable in a short time, and to review the results at the end of each cycle. The review period actually becomes a time to slow down and make sure the direction is still valid or to correct it, before speeding off again. The pace: Go fast, pause to reflect, go fast again. Going fast all the time results in wrecks.

Software inspections and postmortems, unlike CFG reviews, have been used successfully for decades and should be part of every software developer's repertoire. Because they are well-known, I cover only key aspects of them in the remainder of this chapter. Customer focus-group reviews require more description since they are newer, at least in the context of software development.

Customer focus-group reviews, inspections, and postmortems are not the only types of learning techniques, but they are representative of techniques that support the models and assist learning. They are important because they help development team members learn about the product, the customer, and themselves. They involve a wide range of participants and, accordingly, enhance collaboration as well as learning.

Customer Focus-Group Reviews

Understanding the customer's needs is a key issue with today's management, particularly so in the years since the release of Tom Peters and Robert Waterman's book *In Search of Excellence* (Peters82), but many technical organizations still have not gotten the message. They stress customer involvement in projects but accept involvement within their own context, not the customer's. Getting close to the customer means more than accepting involvement—it means embracing it by living in the customer's world and talking his or her language.

In order to achieve real customer satisfaction, software developers must strive to truly think about the customer. In many internal IT groups, the attitude that "we need to be in control and manage the corporate data asset" prevails, but customers are not to be controlled—they are to be provided service.

According to Mike Caldwell, director of Footwear Information and Processes for Nike, "You continually hear IT employees complain that they are not close to the business. 'We want to work closer to the business' is a familiar statement. Well, as long as you describe yourself as not being close to the business, you perpetuate that you don't understand the business. . . . I truly believe that I am deeply ingrained in the sports and fitness business and happen to use my IT knowledge and skills to create, develop, and deliver solutions to our problems" (Highsmith98, p. 3).

One way to create, develop, and deliver solutions is to enable software organizations to concentrate not just on product quality, but also on service quality. Customers, whether internal or external, tend to be most influenced by service quality. Studies (Whitely91) have shown that only a small number of dissatisfied customers complain; instead, they just move their business to other providers. Similar studies have shown how critically important it is to handle complaints well.

A Partnership with Customers

Since customer focus-group reviews are intended to improve the relationship between software development groups and customers, we need to examine some of the issues—and the historical perspective—surrounding this relationship.

One particularly interesting attitude about the issue of customer and product is attributed to Stanley Marcus, a founder of the Neiman Marcus department store: Only customers and products are important; it's the customers that must be taken care of, or they don't return. Simple to say, difficult to do.

It is too simplistic to just say there should be better communication between developers and software customers. Better communication and client involvement have been pursued by many in our industry. Techniques such as JAD sessions have helped improve communication to some extent, but the basic issue is one of attitude, of mental model. Rob Thomsett, a refreshingly entertaining Australian project management consultant, breaks the history of today's often devastating developer-to-customer relationship into four allegorical phases: the Dark

"In observing the excellent companies, and specifically the way they interact with customers, what we found most striking was the consistent presence of obsession. This characteristically occurred as a seemingly unjustifiable overcommitment to some form of quality, reliability, or service."
—T. Peters and R. Waterman [1982], p. 157.

Ages, the Age of Tokenism, the Age of Payback, and the Age of Partnership. As I recall it, Thomsett's history goes something like this.

In the Dark Ages, when computer technology was young and hardware was extraordinarily expensive, only a handful of mystics knew how to mollify the computer deity. Customers went to the mystics, hat in hand, to beg for applications. The mystics were in control, and they reveled in their mystic powers. Starting in the 1970's, in the Age of Tokenism when computer usage became more widespread and the expense of maintaining the hardware and applications grew, the power of the mystics began to slip. Feeling this loss of power, the mystics unveiled a new strategy: Let the customers *think* that they have some input, but keep the mystical secrets under wrap. Have customers participate in requirements definition, get them to sign off in blood that the specifications are accurate, let them help with testing data, and best of all, get them to write the dreaded user manuals. The customers went along, having no choice, but their resentment festered. The technicians, although beginning to lose their mystic status, were still firmly in control, but they were also naïve enough to think that the tokenism wasn't understood by their customers.

As the 1980's unfolded, the Age of Payback arrived. Personal computers, local area networks, and, later, client/server systems provided an alternative to mainframe systems. Thoroughly angered by their long-standing shabby treatment by the IT community, customers rushed to develop their own systems. Cost didn't matter. Finally, the customers had control. The number of individual, work-group, and departmental applications mushroomed and integration and efficiency suffered—but customers weren't about to return control to the mystic techies in the IT groups. Some customers even went so far as to outsource the entire IT department! The cry of the customer rang throughout Computer-land: "We've got those @#$# technoids where we want them now!"

In the 1990's, many organizations entered the Age of Partnership, in which mutual respect, collaboration, and shared responsibility were—and still are—key. In this Age, partnership requires several mental-model changes, such as seeking to understand before rushing to judgment and seeking win-win outcomes. Active partnerships are required for adaptive development to be successful. There are too many decisions to make, too many conflicting requirements to resolve, too many plans to adapt, to be successful with non-partnership relationships. The Age of Partnership set the stage for Adaptive Software

Development, which can be used to build partnerships between customers and developers.

Of course, Thomsett's history tells only an allegory—or does it? In fact, some of today's companies may find themselves way behind the times, and the decade in which a particular company would have entered or left one of Thomsett's Ages varied. Even in the last months of the 1990's, some firms were still locked in the Age of Tokenism or the Age of Payback. A significant requirement for progressing to the Age of Partnership is the ability to understand each party's perspective. Many developers lament that customers just do not understand them or what they do. Next time you are working with a customer, try an experiment: Every time the thought "*you* don't understand" passes through your head or is expressed by someone on your project team, ask, "What is it *I* don't understand about the customer?" Approaching the customer from an attitude of learning, rather than of teaching, pays big benefits. Often in customer focus groups, the developer's first reaction to a criticism from or a change request by the customer is to *explain* or *defend*. Rather than explaining or defending, try another question, one that begins, "Okay, I think I understand part of your point, but help me understand this other part." By seeking understanding, we validate another's contribution. In validating the contribution of others, we open them to listening to us.

Two observations may help developers understand the Speculate component of the Adaptive Development Life Cycle, enabling them to communicate better with their customers:

> **It is impossible to precisely specify software requirements.**

> **The application itself is the only acceptable deliverable on which customers can base an evaluation.**

Since complex projects are nonlinear and can be unpredictable, the belief that one can accomplish precise, complete requirements definition at the beginning of a project is a myth. Sometimes, our technically oriented analytical personalities would have us specify every detail up front, but it is difficult to do so for two simple reasons. First, the requirements constantly change in extreme environments; and second, even when the requirements *are* known, they can be too easily misinterpreted, in part because of the ambiguity of language (Gause89, pp. 14–33).

The issue of the ambiguity of language struck me as especially crucial during a recent project postmortem in which the client commented, "The problem was miscommunication. We often interpreted the spec in radically different ways." My client's observation about the difficulty in specifying requirements did not suggest a wish to abandon trying to define requirements, but it did suggest that any requirements process is doomed to much less accuracy than we probably expect, and that we therefore have to put a process in place to learn what adaptations are required over time.

If we know misinterpretations of requirements definitions are inevitable, how can we best detect problems? We must communicate through a medium that customers can understand, and what they understand best is not data flow diagrams or data models or even object models—but *the application itself*. Consequently, the primary vehicle for confirming requirements is most likely to be iterative versions of the application itself.

JAD sessions have been touted for years as an important technique for gathering specifications for software applications. They engage customers and developers in a common dialogue and speed the development process, but the process does not provide timely *feedback*. Customers attend JAD sessions, usually get documented results, usually feel pretty good about the sessions, but *product results* are still ephemeral. Without the benefit of an adaptive development cycle to deliver partial products, the benefits of JAD are quickly dissipated. JAD sessions provide *feed-forward* information. They need to be matched with appropriate *feedback* sessions.

Objectives of CFG Reviews

A customer focus-group review brings together developers and customers in a facilitated session, the objective of which is to document requested changes to the reviewed application. Customer focus-group sessions are the single most important technique for developers to get *close to the customer* and thereby build good customer relationships.

Like JAD sessions, customer focus groups are special meetings with specific objectives:

- Each session is a review of the product itself, not of documents about the product.

- Each session is designed to help participants find and record customer change requests.

- Each session requires that participants adopt roles, such as that of facilitator.

- Each session stipulates that developers act as if they are behind one-way glass—that is, in general, they are to be seen not heard.

In a well-run focus-group review, everything is a success! Whether comments are positive or negative, something is learned about the product.

For a variety of projects, focus groups have proven remarkably useful. Customers like the fast turnaround; they evaluate a product rather than some indecipherable diagram; their comments are taken seriously; and they invariably become what developers have wanted them to be—involved!

Focus groups have been used by marketing organizations for years to uncover new product requirements and refine products to meet customers' needs. In recent years, PC software vendors have made extensive use of *usability* labs to gain similar insight into how customers use their products. Focus groups have shown that no matter how dedicated the developers are, customers often don't react to products in the way developers think they will, even after extensive initial written specifications. A CFG is possibly the best technique available for building a customer's sense of ownership in the application and confidence in the development team.

Despite their simplicity, customer focus-group reviews are extremely powerful, like the project data sheet. But while the value of the project data sheet seems to be intuitive—evoking comments like, "Yeah, that looks like a pretty neat idea!"—the value of customer focus groups is more subtle. Focus groups build commitment to the project from customers who get an opportunity to examine the product during development. Customers not only learn the application, they become product champions.

Focus groups encourage a collaborative environment between customers and developers by building a common *context* through which to bring diverse information into the product. They

- focus on business results, not technical means

- help developers learn what *close to the customer* means

- provide a forum where the customers feel that they are on a level playing field with the technical team

- increase customer involvement and provide a context more closely understood by the customer

- help participants explore fuzzy areas

Focus groups also facilitate the cultural change in moving customer/developer relationships from the ages of Tokenism and Payback to the Age of Partnership. JAD sessions elicit information from customers on an application's requirements, while focus groups demonstrate product results. Asking a customer about requirements and showing him or her a product version create a different dynamic. The most powerful aspect of focus groups is that they become catalysts for changing an organization's culture. Telling developers they should listen to customers is one thing; having a definitive practice established to facilitate listening is another.

A CFG should have a clear objective and an agreed-upon agenda and should focus on demonstrating business scenarios. A business scenario might be stated as "Show that the system has the ability to create appropriate access rights between people and documents" or "Show that the system has the ability to produce an access-rights summary report." This approach is in contrast to one demonstrating how to change background colors or how to navigate around the application. Customers are interested in solving business problems. Successful focus groups address business scenarios.

Customers can also come from technical areas. On one data architecture project, the team I was working with identified a group of technical customers and a group of business customers. We devised a different focus-group session for each. The final product of the project team was a data architecture document, not a software application. Reviewing the results in draft form over three cycles led to an architectural plan whose implementation started within six weeks after the plan was issued. In large part because of the customer focus groups and the collaboration with both technical and business customers, we could deliver implementation support with the plan.

Customer focus groups are not prototype sessions. Prototype sessions, in which customers and developers sit down together and review or develop applications, should be part of any requirements gathering or development effort. Customer focus groups are more formal, they review the application at the end of a major development cycle, and they involve a wider audience than prototype sessions.

The customer focus-group review process is the same as that for any well-run meeting, involving preparing for the session, conducting the session, and evaluating the final results.

Preparing for the CFG Session

Although the goal of demonstrating application capabilities and recording change requests is basic to every CFG, the objective of each focus group may be different and, therefore, should be explicitly stated. Three key question areas must be considered:

- What is the objective of the session?

- What does each participant need to get out of the session?

- What will be produced, who will use it, and how will each deliverable be used?

The facilitator has primary responsibility for focus-group preparation. It is important that all participants understand their role in the process and that all relevant information be gathered and sent to participants ahead of time.

Preparation Tips

The facilitator and project manager need to provide participants with introductory material, the session schedule, the agenda, their duties, and an explanation of how the focus group fits into the development process. Other tips for sessions follow:

- Include a list of the components to be reviewed in the session as part of the material sent to the participants.

- Limit focus-group sessions to a dozen participants, holding multiple sessions if necessary.

- Ask the facilitator to sit through a dress rehearsal prior to the session.

- Assure that all work on the product cease after the dress rehearsal.

- Decide how change requests are to be recorded.

The facilitator and the project manager must also make sure that there is an adequate cross section of customer reviewers. It is easy to have the system biased toward one group at the expense of another. Focus groups with the wrong participants will obviously skew the results. The participants selected should be drawn from both the vertical levels of management and the horizontal strata of actual system users. In order to insure this cross section, several teams I have worked with in North America have conducted focus groups in Europe for applications impacting their company in multiple global locations.

Conducting the CFG Session

The facilitator conducts the focus-group session, leading the developers and clients through the planned business scenarios. He or she is responsible for keeping the session on track and moving forward with agenda items. For example, since the focus of cycle 1 is on scope and concept, the facilitator would steer questions about screen layout preferences to later sessions. A sample agenda for a session might include

- a CFG process overview and session objectives

- a short review of the project data sheet, the project objective statement, and the mission-profile matrix

- a demonstration of the product using business scenarios

- a review of any open issues identified during the session

- a summary and review of change requests

Because the primary objective of focus-group sessions is to uncover needed changes to the product, participants should concentrate on *what* changes need to be made, not *how* to implement them. The session should maintain its focus on the product as it is presented. If a missing feature set or missing functional area is identified, the item should be recorded and the session should continue. For the missing functionality, convene a separate JAD session rather than allowing time for immediate development of specifications in the CFG meeting.

As stated in the previous paragraph, the objective of focus-group sessions is to determine what changes customers want to see in the application. The focus-group session is *not* a change-request evaluation period. There may be some discussion between participants about a change, but generally the development team should not offer feasibility evaluations at this point. However, the facilitator should discourage discussion clearly outside the scope of the project, or outside the feature set under review. The most effective comments related to project scope come from customer representatives on the core team. Watching customer staff representatives tell their own management, two or three levels up, that something is out of scope is a real treat. Nothing solidifies the benefit of these sessions for the developers more than having customer team members helping to set project scope boundaries.

During demonstrations of early versions, the developers may need to conduct the interim product demonstrations because the software may be relatively unstable. As the product evolves toward completion, customer core-team members should take a more active role at the keyboard.

Facilitators face a particular challenge with developers. It is hard for developers to sit back and just *listen*. In one session, a developer blurted out, "We can't do that." My response was, "Fine, but let's record it as a customer change request and analyze it later." At the break, the developer came to me and said, "I've figured out a really simple way to do it." The immediate "no way" was a placeholder.

Saying *no* to a client change request can curtail discussion early in a focus-group session. All that many clients need is a couple of negative responses to trigger their latent distrust of developers, and they begin to hold back. A good indicator of a CFG's success is increasing levels of group energy, contribution, and laughter during a session. By letting changes get recorded rather than trying to analyze them, facilitators encourage input, most of which will enhance the project. Unfortunately, cutting off inappropriate changes can result in cutting off positive suggestions also. The session debrief is the time to weed out inappropriate changes.

Conducting Tips

This session makes possible a client review of the product. Tips include

- Limit technical-developer comments to client- or facilitator-requested clarifications.

- Keep in mind that both positive and negative comments about the product provide equally valuable information. (Negative comments may lead to changes that are even more useful than those changes brought about by positive comments.)

- Avoid defensiveness or overexplanation of *why* something was done in a certain way by the developers as such behavior will reduce the usefulness of the session. Checking egos at the door is hard, but critical to success.

Evaluating Focus-Group Results

There are four steps to consider in reviewing a focus-group session: first is evaluation of the session itself; second, consolidation and publication of any documentation developed during the session; third, development of a strategy on how to handle the requested changes; and fourth, assignment of issues raised in the session for resolution. The review of the session is done by the core project team.

The four review steps may actually be done in two meetings. The core team should convene immediately after the focus group has finished to share impressions of how the meeting went and to gather recommendations for improving subsequent sessions. It might also discuss items of an organizational or political nature impacting the project—for example, whether a representative is needed from an additional area or an attendee should be released from further sessions. If held the same day as the focus group, this review meeting should be short, but it is important to record meeting impressions before they are forgotten.

The remaining three steps can be addressed on another day. The main job of the follow-up meeting is to categorize, evaluate, and determine the action to take on each change request recorded during the focus-group meeting. Session debrief is the time to critically examine change requests. Within the confines of the core team, it is appropriate to say, "This is a really dumb request!" Although sometimes, on reflection, a request proves not so dumb as it may first appear.

Customers understand that some requests are not feasible; they just don't want to feel that requests have been summarily rejected. Going back to focus-group participants with a list of 48 change requests, where the team has recommended that 42 be implemented, four be studied, one be deferred until a later cycle, and one be rejected is *always* well received. If customers feel strongly about the rejected changes, the discussion of trade-offs takes on a different flavor. In one such discussion I led several years ago, it was clear that the customer's requested change would add two weeks to the four-month schedule. Having attended two focus group sessions and seen previous change requests implemented, the customer had confidence in the development team's two-week estimate. The customer's response was to authorize the developers to take two additional weeks—the time was viewed as a reasonable trade-off, and the customer was happy to approve the short additional time to get the requested change.

Software Inspections

Of the many sophisticated software engineering techniques implemented during the last fifteen-to-twenty years, the one with the most consistently proven track record for quality and productivity improvement is the technique of software inspections.[1] A NASA study published in 1990 stated that inspections were the single most important quality improvement technique.

"Reviews don't slow projects down, defects slow projects down!"
—G. Weinberg, Notes for "Quality Software Development" Workshop, 1989.

In *Applied Software Measurement* (Jones92), Capers Jones's evaluation of defect-removal techniques shows that requirements reviews have a 40 percent defect-removal efficiency, design reviews have a 55 percent defect-removal efficiency, and code reviews have a 65 percent defect-removal efficiency. Other studies conducted by industry researchers and derived from internal company reports have yielded similar results.

It is interesting to note that the single most effective defect-removal technique—at least among those techniques for which quantitative data are available—for software development is software inspection—a *collaborative practice*. According to Jones's statistics, JAD sessions are also one of the most effective defect-prevention techniques—again, a collaborative practice.

The benefit most often cited for inspections is their value in finding defects, but the more important, and often overlooked, benefit is that inspections accelerate team learning. The general rule of thumb for inspections is that a defect found in six-to-sixteen hours of testing can be found in one-to-two hours of inspection. However, being a part of a team that conducts regular inspections makes producers of deliverables more aware—and they generally can find the defect in two-to-fifteen minutes.

---

[1] Although there are differences between reviews, inspections, and walkthroughs, I use the term "software inspections" here to encompass all three categories.

In spite of all the recognized benefits of inspections, the activity is still only a single technique. Testing finds a different set of errors. Another conclusion we can draw from Jones's statistics is that high levels of defect removal are only reached by a well-considered combination of defect-removal and defect-prevention techniques.

Inspections have numerous benefits. They can be used to test a wide range of deliverables from a strategic plan to a test-case scenario. For most early-life-cycle deliverables, inspections are the only viable quality assurance technique. Inspections also create product consistency through the constant review of team members' work and subsequent incorporation of each other's best practices. Without a good inspection process, documents such as coding guidelines are lifeless. Inspections provide short-term feedback for products, and longer-term feedback on processes and performance. Finally, inspections provide a method to validate non-testable or hard-to-test capabilities. For example, without expending considerable effort, developers find it difficult to *test* for maintainability of an application.

Despite the proven benefit of inspections, there are still significant challenges to making them effective in an organization. As with any of the collaborative tools, inspections can build up or tear down relationships. Once inspections become destructive to relationships, the inspection process will not survive in the organization—or if it does survive due to management pressure, it will provide much form, but little content. There are several keys to effective reviews:

- Every inspection should have a definitive objective.

- Inspections should be limited to one-to-two hours.

- Inspections need to be tuned for effectiveness.

- Inspections should be facilitated.

- The inspection itself should be done prior to the meeting.

There are different kinds of inspections, and different objectives for each.[2] Objectives guide participants both in preparation for and in conducting the sessions. Inspections are short, specific, intense meetings. If the inspection session cannot be completed in one-to-two hours, it is either too large or the team is discussing issues rather than inspecting.

---

[2] Some concepts in this section on inspections are loosely adapted from my colleague Lynne Nix's "Software Inspections" seminar, available through Knowledge Structures, Inc., of Sacramento, California.

Tuning inspections for effectiveness involves adjusting the number of people in a session and the time spent in preparing and inspecting. For example, six people in a code inspection is probably too many, whereas three people in a requirements document inspection is probably too few. A minimal set of metrics, such as defects found per hour, can help tune inspections for both effectiveness and efficiency. Checklists greatly increase inspection effectiveness, partly because they help participants know what to look for. Updated, refined checklists are as much a product of inspections as are the reviewed products themselves.

The facilitator must make sure that the participants understand that the actual inspection is done prior to the meeting. The session itself is meant to review the identified potential defects and to record those the author of the product agrees are defects.

Preparing for the Inspection

The facilitator or moderator of an inspection and the product's primary author are responsible for putting together the packet of material for inspection and for preparing the participants. An inspection packet contains, at a minimum, a work product, objectives, and a checklist. A complete packet should contain

- a work product—that is, any deliverable component from the cycle plan

- a list of objectives that describe the *kinds* of defects the inspection is targeting

- checklists and questions that help inspectors identify defects, and that mirror the objectives

- reference material containing related work products used for comparison and consistency purposes, or a design document that might be a reference for a code module

- guidelines representing an organization's best practices for a type of work product, such as a set of guidelines for coding in a particular language or a template for a requirements specification

To prepare for the inspection, the author and the facilitator select the participants, set the schedule, and determine whether any pre-inspection meetings are required.

Conducting the Inspection

The objective of an inspection is to discover defects, not to correct them. One reason inspection processes fail is lack of focus. If inspections become generalized discussion sessions, their effectiveness will plummet. The facilitator's primary job during an inspection is to maintain this focus. If the group wants to discuss solutions or concentrate on an aspect of the project other than finding defects, the facilitator should terminate the inspection, ask the participants to leave the room, and then ask them to reenter for a general discussion meeting. It is critical to the ongoing success of inspections that the group view them as short, focused, productive meetings. By making crystal clear the demarcation between the inspection meeting and the problem-solving meeting (even to the point of the seemingly superfluous act of having attendees leave and return to the room), the facilitator helps all participants to become more aware of specific kinds of group interaction.

Inspection Tips

- Start the session with a round robin of three *positive* comments from everyone about the product in order to set an upbeat tone. Someone invariably brings up something funny.

- If participants haven't prepared adequately for the session, stop the session and reschedule it.

- Insist on specifics, encouraging statements such as, "On page 4, line 34, there appears to be a missing condition."

- The author(s) of the work product have the final decision over whether or not something is a defect.

Evaluating Inspection Results

There are three primary results of an inspection—the defect list, suggested updates to checklists, and any inspection metrics (such as hours spent per defect discovered). The work-product author and the facilitator determine whether the product will need to be reinspected after the corrections have been made. The facilitator is responsible for initiating whatever process is in place for updating checklists.

A few metrics, such as knowing how many pages of documents or lines of code can be inspected per hour, can help the group to tune inspections to be more effective. In addition, knowing the difference between the cost of code-inspection groups consisting of three participants versus four and the number of defects found by different sized groups can have a big impact.

Project Postmortems

Just like fishing stories, climbing stories get longer with repeated telling. Talking about the adventure is sometimes more fun than actually experiencing it—just ask anyone who has spent hours slogging up a glacier anticipating that each successive ridge will be the summit. Climbers recount each and every phase of the endeavor, from the food to the long walk back to the most difficult move. Decisions made along the way are reviewed and analyzed ad nauseum. Nonparticipants (for example, friends or even spouses) can become extremely bored by such debriefings, but these sessions can prove instructive in preparing climbers for the next outing (such as reminding the group, Never let Jim cook again).

Every book on project management advocates project postmortems or retrospectives, yet few companies do them at all, much less do them well. Postmortems don't have to be brutal, but they are critically important to extreme projects. In fact, with extreme projects, there should be mini-postmortems at the end of each adaptive cycle. These cycle reviews should both reflect progress-to-date and anticipate the next cycle. Finding and correcting mistakes early is preferable to waiting until the end of the project.

Postmortems must be supportive of staff efforts by maintaining an underlying attitude that recognizes that everyone does the best job he or she can do under the circumstances. It is easy to Monday-morning-quarterback both football games and development projects. The objective of the postmortem is to help people learn from mistakes and to anticipate the future, not to find fault. The most common reason for postmortems failing in practice is their being used to assign blame rather than to learn. People figure out very quickly which kind of postmortem philosophy is being used.

> **In a blaming environment, postmortems will be full of content-free words.**

Postmortems test the degree of trust and respect between members of a team more than any other type of meeting. While customer focus-group sessions intentionally bring up criticism of the product, facilitators must conduct postmortems so as to avoid any implied criticism of *individuals*. Facilitators strive to avoid such criticism, but it is inevitable that feelings get hurt. Effective teams can move past the hurt feelings and learn from the interaction; however, if trust levels are low, a postmortem can easily drive them lower.

There is a need for project members to continually assess all aspects of the project and to make hard decisions. A complete assessment should look not only at the project's progress, but, from time to time, even at the objectives themselves. Unfortunately, to consider abandoning or significantly altering the original objectives of a project is anathema to most organizations. Although they usually have some type of project-progress assessment in place, the range of potential actions based on the assessment is narrowly limited to politically acceptable options. An adaptive project leaves room for large alterations in the project's objectives. According to Capers Jones (Jones96), the largest software failures, costing millions of dollars, usually occur toward the end of a project, when the lack of results can no longer be hidden. It is my firm belief that

> **Project teams have to face defeat in the end because they don't allow themselves to *seriously* consider defeat (and, therefore, altering course) in the middle.**

Preparing for the Postmortem

Preparation for a postmortem session should be similar to preparation for other collaborative meetings: Develop an objective, decide on attendees, collect project information, prepare the participants, and manage the logistics. In addition to the preparation activities already covered for other types of collaborative sessions, here are some specific suggestions for postmortems:

- Conduct the session while the experience is still fresh.

- Encourage participation by all team members, including managers and leaders.

- Use an outside facilitator.

- Discuss successes *and* problems.

The facilitator should send a notice to all participants before the session, stating the objectives, agenda, and each participant's role; further, he or she should encourage each prospective attendee to consider the successes and challenges of the work to date. If the team has not conducted a postmortem previously, a sample postmortem document or, at the least, an outline of its contents should be included with the preparation material.

The facilitator and project manager should schedule the session while the project is still fresh in people's minds (certainly within the first month after completion). Cycle postmortems should be held within days of the cycle focus-group session. Any improvements for the next cycle need to be put in place quickly.

Since managers and team leaders are highly involved in the project, significant information—and trust!—is lost if they are excluded from the postmortem. Trust is especially important. If there is not enough trust in the group to include managers, there is not enough trust to have productive postmortems. However, because of the emotional stress postmortems can generate and the difficulty in facilitating the investigation of one's own foibles, it is very important that the sessions be facilitated by a project outsider. Having a facilitator enables the team members, including leaders and managers, to concentrate on the content, not on the meeting process.

People have a tendency to be overly critical in postmortems. The facilitator can counteract this potential negative bias by asking participants to develop a list of three project successes and three project problems before they attend the session. Thinking about and discussing specific successes and problems helps maintain a balanced perspective during the session. This exercise also helps counteract the downside of conducting postmortems within a few weeks of the project's end when problems are often the most vividly remembered items.

Conducting the Postmortem Session

The main item on a postmortem session agenda should be to analyze the overall success of the project using the mission statements and the project data sheet as guides. The discussion should revolve around three questions asked about each project dimension. The questions are: What worked well? What problems were encountered? and What are areas for future improvement?

A simple, relatively unstructured postmortem might just pose the three questions. However, it is usually beneficial to identify project dimensions and ask the questions about each area. Examples of these dimensions are

- the product mission profile (scope, schedule, resources, and defects)

- project management (planning, scheduling, risk management, status reporting, resource planning)

- adaptation (team and process performance)

- collaboration (team jell and compatibility)

- learning (both about ourselves and about our mental models)

- balance (flexible and rigorous process usage)

Evaluating Postmortem Results

Postmortems should be documented, and the documents widely distributed. It is not necessary for every project team to learn from the same mistakes—learning from someone else's mistakes is more productive. One measure of the effectiveness of postmortems in an organization is to determine how widely they are distributed to and used by other teams. A narrow distribution indicates narrow learning—in fact, narrow distribution usually indicates no learning whatsoever.

Summary

➤ How many *team* learning techniques (JAD sessions, focus groups, inspections, or postmortems) are effectively used in your organization? You now should have a feeling for how well your organization learns.

➤ There are at least two types of learning—learning about things and learning about ourselves, as illustrated by the difference in training for a climbing style that one already embraces, and changing one's mental model to accept a new climbing style.

➤ Success is not about what we know; it is about how we learn.

➤ Success is determined by the adequacy of the feedback, not the feed-forward.

➤ Senge's learning model contains five key components: systems thinking, team learning, mental models, shared vision, and personal mastery.

➤ Innovative learning is about understanding the exception conditions to our general mental models.

➤ Silver bullets are new mental models for which the exception conditions are not yet understood.

➤ Big changes often lead to failure. Little changes often lead to big changes.

➤ Learning techniques embody learning concepts. At the core of the techniques is facing our overestimation of our own understanding.

➤ Learning techniques must help produce both better products and better collaborative relationships. Failure to do either will render the practice hollow and ineffective.

➤ "Close to the customer" has become a slogan. Many organizations espouse slogans but never follow up with either the required cultural (mental model) changes or the techniques to make the slogan real.

➤ Seeking to understand, then to be understood, combined with a practice like customer focus groups, helps make "close to the customer" a reality.

➤ One key to treating customers as collaborators is utilizing familiar *contexts*. Hence, the belief that the application itself is the only acceptable deliverable on which customers can base an evaluation.

➤ Software inspections are one of the most widely measured and proven software engineering techniques. They are intended to uncover defects, but just as importantly, they serve to enhance technical team learning.

➤ Postmortems provide a respite that allows for evaluation of successes and challenges, and an opportunity to redirect during a project and reflect afterwards. Although simple in concept, they are one of the most important and most difficult of the learning techniques.

➤ In a blaming environment, postmortems will be full of content-free words.

Part 3

Part 3

CHAPTER 7
Why Even Good Managers Cause Projects to Fail

O n the desk in my office sits an entirely self-contained, enclosed, living ecosystem. Five inches high, this microcosm of life contains algae, very small shrimp and snails, and multitudes of microscopic bacteria, all living by exchanging *stuff* with each other and by converting light to biochemical energy. Next to this living ecosystem, on my computer monitor, digital beings (Biots) exchange digital *stuff*. They live, eat (each other, of course), mate, give birth, evolve into new organisms, and die—a sea of artificial life created in silicon by one of the many Artificial Life programs available.

I use these two visions of life, one real and one simulated, as constant reminders to myself to think about all business organizations as if they were organic ecosystems. Thinking of a feature team, a larger product team, or an entire company as an ecosystem helps me to understand how limited one's actions are to influence the forces that propel the economic landscape. When thinking about how to manage in complex situations, I find that the concept of a living ecosystem provides a profound cultural perspective.

Biologists can't predict what will happen when an exotic species is introduced into an ecosystem; neither can business organizations. Other species will react to the exotic in innumerable ways, some with short-term impact, others with long-term—for example, early European settlers introduced rabbits to New Zealand as a food source but, with no indigenous predators, they have now become so abundant as to be a scourge to farmers. An ecosystem suggests the presence of living things and their interaction. Market share, financial analysis, data warehouse—all bring to mind things without life. Animate or inanimate, each demands a different perspective. To think intelligently about a biological ecosystem or a business system, one must have a solid understanding of adaptation and evolution—one must focus on the dynamic rather than the static.

A scientific phenomenon in which an ecosystem experiences dynamic or sudden change (caused by an outside force) is called a "punctuated equilibrium." As in the case of an outside force's effect on the dinosaurs, some species adapt, others don't. There is a similar concept in business, most commonly found in high-technology businesses, called "disruptive technology." Like punctuated equilibrium in biological ecosystems, disruptive technologies allow some firms to prosper while causing others to die. Disruptive technologies change markets in ways that leave existing management strategies ineffective. Disruptive technologies cause even good managers to fail.

Disruptive Technologies

Our assumption that effective managers deliver successful products and ineffective managers cause projects to fail is wrong. Clayton Christensen paints a different picture of management success and failure in *The Innovator's Dilemma: When New Technologies Cause Great Firms to Fail* (Christensen97). Christensen identifies two types of technologies—sustaining and disruptive—the second of which alters the basis of competition.

Changes in sustaining technology can be significant, but the criterion that separates sustaining from disruptive technologies is the impact sustaining technology has on business value. Sustaining-technology changes provide the customer with better, cheaper, faster versions of the same product. Disruptive technologies offer a new product with a different value proposition. For example, a technology that increased the capacity of an 8-inch disk drive would be a sustaining technology—same product, greater capacity. A 5¼-inch disk drive (smaller, less capacity, higher price) offers a new value proposition—

smaller size. However, and this is a critical point, the new proposition (size) may only be of value in a new market. If the PC market had fizzled, and the mini-computer market had continued to grow, the smaller-size disk drive might not have been disruptive at all because it would never have gained enough market share to "disrupt" the 8-inch drive market.

Firmly established in the new market for PC disk drives, the smaller drives' performance curve increased more sharply than that of the larger drives. Once the performance of the 5$\frac{1}{4}$-inch drives began exceeding that of the 8-inch drives—smaller *and* better performance—the 8-inch drive market shriveled quickly.

Likewise, disruptive management practices are only relevant when the environment changes sufficiently such that traditional practices no longer produce the necessary level of success. For example, the impetus to use adaptive management practices will come only in environments in which speed and change cause optimizing practice to become ineffective.

In studying the history of companies in the disk-drive industry during the years between 1976 and 1996, Christensen found that in every case, without exception, established firms led the market in introducing sustaining technologies while new entrants dominated the marketplace in disruptive technologies (that is, by changing the disk-drive size). In reality, two years after the first 5$\frac{1}{4}$-inch drives were introduced, 80 percent of the manufacturers were entrant firms—few of the 8-inch-drive makers survived the transition to the new 5$\frac{1}{4}$-inch market. And, while established firms who were latecomers to major sustaining technology initiatives (such as thin-film drive heads) didn't suffer revenue problems, the difference between late versus leading entrants into 5$\frac{1}{4}$-inch drive disruptive technology was an average of $65 million versus $1.9 billion in revenue (Christensen97)!

The most thought-provoking of Christensen's conclusions is that ineffective management was not the cause of failure. By every traditional measure of a good manager—that is, ones who listen to customers, pursue higher-margin business, invest prudently in manufacturing capacity, and generate lofty financial returns—good managers were the ones that failed. Competitor's attacks from below (that is, from lower-cost, fewer-feature products) were not seen as threats until it was too late.

Just as 5$\frac{1}{4}$-inch drives replaced 8-inch ones, personal computers ousted mini-computers, and hand-held calculators replaced slide rules, disruptive management practices such as Adaptive Software Development, in my opinion, will replace more traditional sustaining practices. Management practices in high-tech firms are different from those in more traditional industries. They are different because high-tech firms respond to the marketplace with what works—and what works in fast, uncertain, changing conditions is different from what works in slower, more certain, relatively stable ones.

Understanding why even good managers fail provides a context from which to view the next four chapters. We have seen, in Chapters 3 through 6, the Adaptive Development Model Life Cycle and the components of Speculate, Collaborate, and Learn. This chapter, and the three following, develop the concept of the Adaptive Management Model, specifically, the transition from a Command–Control view of management to a Leadership–Collaboration approach. The balance of this chapter describes *why* a new management model is imperative and characterizes the market ecosystem in which software development teams must operate. Chapter 8 describes adaptive management in more detail, in particular its culture. Chapters 9 and 10 then explain techniques for scaling adaptive development to larger projects.

High Change

"Complex systems tend to locate themselves at a place we call 'the edge of chaos.' We imagine the edge of chaos as a place where there is enough innovation to keep a living system vibrant, and enough stability to keep it from collapsing into anarchy. ...Too much change is as destructive as too little. Only at the edge of chaos can complex systems flourish."
—M. Crichton [1995], pp. 2–3.

Good managers fail when the rate of change disrupts their ability to cope by means of their usual management practices.

High speed is easy; high change is the real challenge.

High change, when fueled by speed and other external factors, is both hard to accomplish and dangerous. How challenging can be seen as analogous with a high-speed mountaineering gamble taken by an acquaintance of mine several years ago. Tired of the long walk off Denali after having spent three weeks on the mountain, this climber viewed the walk down the Kahiltna Glacier as boring and just wanted to get down. He decided on an exhilarating descent, climbed aboard the plastic sled he used to haul supplies up the mountain, and off he went. Unbelievably lucky, he made it. The Kahiltna is a crevasse-ridden glacier; it is not a place to travel un-roped. Even within roped parties, climbers often find crevasse falls a part of the Denali experience. Rocketing down the glacier on a sled is definitely a high-speed gamble.

Most management strategies are geared either to reduce the number of changes or to control changes. While these strategies are useful, it is more useful to embrace change than to try to control it. By embracing change, management may be able to develop an ability to absorb it. In increasing-returns markets, and especially in fast-paced markets, an organization that can absorb change, learn from it, and deliver in the face of it, has the advantage. Although "change" often has a negative connotation, in unstable environments we should view change as positive, as something providing both an opportunity to learn and an opportunity to get ahead of the competition.

Unfortunately, most change-management processes still reflect the predominance of imposed-order thinking. Traditional project management continuously seeks to control by planning a future project state and then by reducing variations in time, cost, quality, and feature set when they deviate from the plan. As a result of the fact that most managers wish to retain stability, change management practices are, for the most part, considered to be methods for handling exception conditions. But complex, turbulent environments are characterized by disequilibrium and low levels of predictability. Change is not an exception; it is the norm.

In extreme environments, *equilibrium* is the exception condition.

In thinking about change management, one needs to divide an environment into its naturally complex and potentially orderly segments. Potentially orderly parts of the environment are those that are somewhat predictable, and therefore are amenable to optimizing change-control practices. Naturally complex parts of the environment are not amenable to deterministic practices; to manage them, one must utilize a strategy of change *containment*. In high-change environments, trying to react explicitly to every change consumes too much time. Containing change (covered in more depth in Chapter 11) is a limiting strategy, which enables team members to react to some changes without the guidance of explicit documentation or specific change-control procedures.

One example of a generally inappropriate change-control strategy is *freezing* of product specifications. This strategy effectively says, "We just don't recognize new changes." A freeze strategy blocks new input, refuses to recognize reality, and retreats into the comfort of order and control. The requirements specification process is nondeterministic—it addresses a naturally disorderly part of the environment. Requirements changes need to be contained, not controlled. A contain strategy might advocate that developers change the code and move on, whereas a control strategy would necessitate careful documentation updates to maintain requirements traceability. Of course, to get a product into the customers' hands, every product development team must at some point finalize the code and ship. But just as in a high-speed Formula One race, the last one to ease off the throttle while careening into a corner usually comes out first—or crashes.

Requirements changes made necessary because of new customer requests or in anticipation of evolving customer needs, changing technology, or competitors' actions are the most difficult changes to manage. For example, the very day in the Fall of 1995 that Microsoft launched Microsoft Network (MSN), it was rendered obsolete by the Internet explosion. External changes of that ilk cannot be predicted. Even if there were time to employ deterministic change-control practices, they would ultimately fail, for unpredictable problems don't succumb to deterministic solutions. The best weapons for managing naturally disorderly changes are emergent solutions, generated by collaborative teams.

Change can impact hundreds of product components, requirements documents, test plans, user documents, product plans, ideas, sketches, doodles, decisions, and designs. Are all of these to be controlled? Of course the answer is no. Change management can run the gamut from establishing elaborate software change-management systems to ignoring the changes.

In complex environments, change needs to reflect normalcy and learning rather than exceptions and control. In order to illuminate this different mental model, the term "difference" management might be substituted for the term "change" management. There can be *differences* between predicted and actual, with each having an equal probability of being wrong. Whereas changes are usually considered things to be reduced or eliminated, *differences just are*.

If unbridled change leads to chaos and perfect predictability leads to stability and stagnation, then our search must be for a way to contain or bound change in order to remain in the transition zone at the edge of chaos. We must set boundaries—not predict results—and let self-organization respond.

Thus, change management—traditionally defined as "exception processing"—goes away. Change management becomes an integral part of adaptive management, particularly the collaboration portion. The practices and tools needed for effective collaboration are in fact those needed for managing continuous change! In both cases, the trick is to manage the difference between information we already have and newly received information. Balancing at the edge requires balancing between change control and change containment. Whereas change control is formal and rigorous, change containment must be less formal and more flexible.

Altering one's approach to change, whether by acquiescing to periodic plan alterations or by accepting change as the normal state, is necessary if one is to succeed in high-change environments. It is an essential component in creating an *adaptive* culture.

No Silver Bullet

Good managers fail when they attempt to use silver-bullet solutions to complex problems. Jerry Weinberg writes in *Quality Software Management, Volume 3: Congruent Action* that

> *There is no silver bullet, but sometimes there is a Lone Ranger.*
> —G. Weinberg [1994], p. 1.

Weinberg explains exactly what he means by this, but the reason I like the sentence so much is that it highlights the dichotomy between an emphasis on things (bullets) and an emphasis on people (even a fictional character like the Lone Ranger). Let me take my interpretation one step further and amend Weinberg's line:

> **There is no silver bullet, but there are Lone Rangers who have arsenals of bullets for different situations.**

For most managers, the hard part is understanding the different types of bullets and the situations in which each is most likely to succeed.

Ultimately, software projects don't succeed because of techniques; they succeed because of ideas, individual responsibility, basic principles shared by team members, and collaborative interaction. There is no silver bullet—a truth that applies to mountain climbing as aptly as it applies to software development. A mountain climber knows that there are many different mountains, and that each has more than one route to the top. There are many climbers, each with unique skills and experience. Although better tools have greatly advanced the sport of climbing over the last fifty years, it is still the effort of the individual in combination with the climbing team that makes the difference between their reaching the summit or having to retreat.

Why, then, do so many development organizations *say* there are no silver bullets for software, and yet at the same time *act* as if there are? This idea of a silver bullet and the continuing denial that one exists is puzzling. What does "no silver bullet" really mean? The easy answer—at a superficial level—usually is that a particular tool, method, or concept isn't going to solve all our problems. But this interpretation is too simplistic; the essential next question should be asked: "What about the no-silver-bullet concept has caused it to be so widely discussed in software engineering literature?"

One reason the silver-bullet concept remains a topic in our literature and a tickler in our psyches is that, in an era of constant change, uncertainty, growing complexity, increasing speed, and overwhelming competitive pressure, we crave something simple upon which to build a battlement against the forces of chaos around us. We know a silver bullet doesn't exist, but we want one anyway. Whether the silver bullet is in the guise of Total Quality Management (TQM) or object technology, its existence injects the *illusion* of stability into our chaotic, anxiety-ridden world.

Tremendous amounts of time and money have been poured into this quest for stability and the elimination of unpredictability in businesses, in general, and in software development, in particular. Business Process Reengineering was one such silver-bullet effort for which the results never matched the original expectations. Studies reported in the mid-nineties have shown that few BPR projects were successful (for examples, see Tapscott96). Despite everyone's belief in them, why have so many silver bullets been such a dismal failure?

One answer comes from Ralph Stacey, professor of management and director of the Complexity and Management Center at the University of Hertfordshire in Great Britain, who provides a new perspective on the issue of silver-bullet failure. In his 1996 book, *Complexity and Creativity in Organization*, Stacey identifies an essential mental shift needed to understand the failure. The paradox is that the silver-bullet solutions to business problems become the source of the very failures that they were designed to prevent.

Stacey contends that current silver bullets—such as TQM, BPR, and the CMM—create a vicious cycle of problem definition and solution. I call this cycle the *optimization paradox*, and it follows the stages outlined below:

1. Some problem (for example, the need to reduce costs, increase quality, or reduce cycle time for some business function) is articulated by business managers.

2. The solution (for example, BPR) is based on a deterministic organizational strategy in which the company analyzes the situation and then designs a solution based on an assumption that the business environment is predictable and stable.

3. The business environment of course turns out to be ambiguous and unpredictable; therefore, the "reengineered processes" are faced with unanticipated variations that adversely impact the success measures. The dichotomy between the real and ideal worlds increases management and staff anxiety.

4. To correct this new problem, management selects models of business strategy that are prediction- and control-oriented, the assumption being that better prediction and control methods are needed; hence the reliance on well-articulated, engineering solutions such as BPR as a vehicle for imposed order.

5. The business may respond to the silver bullet favorably for a short period, but because rigid processes exclude data outside "acceptable" limits, critical feedback is denied. Since the process "can't be faulty," failure to meet predetermined success measures is blamed on poor execution and therefore "motivation" and "control" efforts are redoubled.

6. The failure to achieve desired results is identified as the next "problem" and a new silver bullet is proposed.

The next silver bullet starts the cycle over again. The fundamental reality is that the business environment is ambiguous and unpredictable, but this reality is never seriously acknowledged. The statement, "I don't know how much it will cost or how long it will take or exactly what we can produce" is not acceptable from someone who wants to be a "real" manager. Managers may pay lip service to constant change, but they are not willing to adjust their fundamental views of management to accommodate the real impact of that change. On one hand, they acknowledge the need for change, but on the other, they still think they can beat the odds and succeed with their "plans" in the face of change—denial in action.

In this problem–solution–problem cycle, millions of dollars are spent for ever-decreasing results. Organizations in this cycle experience a lot of change, but little learning. They are trying to optimize the non-optimizable. Stacey contends that breaking the silver-bullet syndrome depends upon breaking the linear, imposed-order mental model of business and substituting a belief in emergent order.

Most managers use silver bullets in the hope that they will provide prescriptive answers to complex situations. Most soon discover, however, that these prescriptive, optimizing answers get bogged down in metrics, procedures, forms, and slogans—and use of the chosen solution often lasts only until the next fad comes along. Silver bullets can be useful, but they are insufficient. They deal best with the stable parts of the environment, not with the rapidly growing complex portion. In complex environments, use of TQM or BPR or even CMM practices may seem to be the answer, but they are not sufficient for success.

"[W]e respond to the fact that situations are uncertain and conflictual with a rigid injunction that people be more certain and more consensual. . . ."
—R. Stacey [1996], p. 7.

Are Organizations True Complex Adaptive Systems?

Good managers fail because their management practices were developed to solve complicated problems, but their problems are much more than complicated; they are complex.

Altering fundamental assumptions about how organizations and projects should be managed is a difficult assignment. By using the theory of complex adaptive systems to help convince managers of better solutions for their problems, we move closer to accomplishing our goal, but it also forces us to analyze whether or not organizations are, in fact, complex adaptive systems, or whether CAS is merely a useful metaphor.

Ralph Stacey's compelling argument that organizations *are* complex adaptive systems begins with an analysis of the properties of non-linear deterministic networks, which underlie much of the chemical and physical world. According to Stacey, analysis of these physical, deterministic systems has shown that self-organization and emergence occur under certain conditions. Stacey maintains that even if organizations were as deterministic as these physical realms, "[W]e can get such a system to do what we want only if what we want is an endless repetition of what it has already done" (Stacey96, p. 71). Predictability, even in deterministic systems, isn't as predictable as we may think.

Stacey continues his point by detailing the characteristics of complex adaptive systems, which he differentiates from deterministic systems by the addition of a purpose. A deterministic system (for example, a chemical reaction) has no purpose or goal, but even the simplest living system has a goal (to live or to reproduce, for example) and a set of internal rules of behavior it uses to achieve that purpose. Stacey identifies the general features of complex adaptive systems that are shared by human systems. These were first identified in Chapter 1, naming a complex adaptive system as an ensemble of independent agents,

- who interact to create an ecosystem,

- whose interaction is defined by the exchange of information,

- whose individual actions are based on some system of internal rules,

- who self-organize in nonlinear ways to produce emergent results,

- who exhibit characteristics of both order and chaos, and

- who evolve over time.

The final building block in Stacey's carefully constructed case is his identification of the characteristics that differentiate human systems from other CAS, and his analysis of whether those differences invalidate the comparison. He contends that the primary difference is a human's internal structure—consciousness, emotions, and self-awareness—which makes humans and their organizations complex and, therefore, good examples of complex adaptive systems.

Stacey concludes that organizations and the individuals who compose them are, in fact, complex adaptive systems. What we learn about CAS topics such as emergence, nonlinear networks, self-organization, interactions in ecosystems, and adaptability is, therefore, perfectly applicable to how we manage complex adaptive organizations. The Adaptive Management Model and, in fact, Adaptive Software Development in its entirety are built upon this premise.

Requisite Variety

Good managers fail because they don't understand the need in complex situations to know their options and to be flexible—what W.R. Ashby labeled "requisite variety" in his 1964 landmark book *An Introduction to Cybernetics*.

One of the greatest challenges that organizations face is how to establish development frameworks that are flexible enough to allow staff and managers to select the right silver bullets, at the right time, for the right problems. This flexibility comes from the application of sufficient variety to handle complex environments. A manager needs a variety of methods and tools to handle a variety of problems. The more problems, the more complex the problems, the more variety a manager needs in order to solve the problems.

In high-speed, high-change environments, the path to success is through exceptional personal effectiveness from both development staff *and* management. One of the reasons why managing extreme projects is so difficult is that managers must have a large number of coping mechanisms to combat variety in the environment. In low-change circumstances, managers can survive with a brief set of specific rules—if this happens, do X; otherwise, do Y. In high-change circumstances, the range of responses is much more demanding, requiring judgment, juggling, thought, and swift action.

The need for creativity, flexibility, teamwork, collaboration, and fun exists on all projects. Typically, however, as projects move from the 1,000 function-point size into the realm of 10,000 function points and beyond, most organizations revert to discipline, bureaucracy, and strict process control as if customer feedback, iterative learning, and excitement were no longer important.

Requisite variety is the antithesis of silver bullets. Silver-bullet solutions are targeted toward a predictable, optimized world—a world diminishing in size. Extreme environments require adaptive solutions—those with the requisite variety to address challenges from many directions.

Project Ecosystems

Good managers fail because they don't understand the business ecosystem in which their projects must live.

Software projects, whether within a package software company or an internal IT organization, operate within the ecosystem of the company or larger organization of which they are a part. An entire organization, in turn, operates within wider markets made up of competing organizations, which in their own right are also a kind of ecosystem. The product mission profile (defined in Chapter 3) focuses on the key product characteristics. To dominate a market in today's world, a product has to be the best at something specific, not just good at a variety of somethings. In turn, a product's profile—its scope, schedule, resources, and defects—dictates development strategies. But in order to understand the reasons for a particular product profile, one must have a clear understanding of the organization's focus in the marketplace, that is, of how the firm fits within its competitive environment or ecosystem. In ecosystems, every action of one species is met with a reaction from another. Similarly, project teams and companies rarely control their own destinies; they operate within constantly evolving ecosystems. This wider perspective is often lost in the whirlwind of learning new technology and in the hectic pace of development, but it is crucial to the success of a product-development team.

My purpose in this section is not to investigate the fields of marketing and business strategy in depth, but to illustrate how these two aspects of an organization's ecosystem impact its software development strategies. The following explores turbulent markets by examining two current strategic analysis methods: value disciplines and tornado marketing. The description of each of these methods is followed by an analysis of their implications for software development teams.

Value Disciplines

Michael Treacy and Fred Wiersema make the case in *Discipline of Market Leaders* (Treacy95) that a company must "pick a dimension of value on which to stake its market reputation." They define three value disciplines: *operational excellence*, which provides the lowest total-cost-of-product ownership; *product leadership*, which produces the best product; and *customer intimacy*, which provides the best total solution to the customer's problem.

Treacy and Wiersema's premise is that no company can excel in all dimensions; in fact, customers won't let it. Customers understand a company's value proposal inherently: For example, they go to Wal-

Mart for best price and Nordstrom's for best total solution (defined in this case as great service). Customers understand that they will not get both best price and best service under one roof. They expect to pay more for service. They know where to go for each value proposal they seek. They know what they want, and they want more of it—whether it's low price or product excellence. Companies with mixed messages don't lead markets.

Having a value focus doesn't mean ignoring the other dimensions—they too must be satisfactory. If Nordstrom's prices were to get too far out of line, their claim to best-total-solution would lose some validity. Likewise, if Wal-Mart's service were atrocious, customers wouldn't find their low prices so appealing.

Treacy and Wiersema's concept of picking a value discipline necessitates committing the entire company to an operating mode to support it. For example, operationally excellent firms are characterized by discipline, Command–Control management styles, conformance to rules with little variety, and highly integrated, low-cost transaction systems—they streamline operations, reduce variety, and relentlessly drive down costs.

Customer-intimate companies are highly client- and field-driven, and are flexible and responsive. They concentrate skilled staff in the field in order to intimately understand their customers' businesses, and to provide tailored customer information systems. Their emphasis is on creating deep, long-term customer relationships.

Product-leadership companies aggressively experiment, attack competitors' products, encourage risk, problem-solve voraciously, discourage bureaucracy, and foster information systems that actively encourage the cooperation and knowledge-sharing developers need for product innovation.

An IT software project team, supporting a product manager in a product-leadership company, using a by-the-book SEI-defined process strategy, is mixing water and oil. Any talk of integration, low maintenance, defect-free code, and architectural elegance will fall on deaf ears. Just as the product manager's customers are looking for innovative, leading-edge products, the product manager is looking for innovative, fast response from his or her information-support services.

In contrast, using an Adaptive Software Development strategy to support a transportation and distribution manager in an operationally excellent company probably will not work either. In a culture that doesn't like mistakes, stresses conformance to rules, appreciates low-cost alternatives, and thrives on integration, little *real* support for practices (and underlying mental models) that deliver software at high speed exists.

There are two important observations to be drawn from this idea of focusing on a single dimension of value. First, one must understand the business environment and align software development strategy with business strategy. Making the business value visible to the software development team is an important step in the alignment process. Second, if focusing on a single value dimension is necessary for a company to be a market leader, it is logical to assume that a software product or project similarly should have a single focal point. A product needs to be *good* at a lot of things, but it needs to *excel* in a single dimension. This idea about the difference between "good enough" and "excellent" was discussed in detail in Chapter 3 and incorporated into the product mission profile.

Tornado Marketing

While Treacy and Wiersema examine general business strategy, others target high-tech product marketing strategy. Since most high-tech companies exist in increasing- rather than decreasing-returns environments, development team members and managers can better focus product development if they understand more about marketing strategies.

In fact, one of the more enlightening periods of my own professional journey was three years spent as a product marketing manager and vice president of Sales and Marketing for a high-tech firm. The experience changed my attitude significantly—so much so that, in seminars, my admonition to developers is to spend time with sales and marketing staffs if they really want to understand their companies, products, and ecosystems.

In adaptive development environments, a sense of shared vision is essential to the collaborative effort needed to produce outstanding results. Part of this shared vision is understanding the overall marketing strategy for the product, where the product lies in its life cycle, and how development strategies need to reflect the product's life cycle stage. Geoffrey Moore's work in the area of high-tech marketing might be called "Technology Adoption Life Cycle With A Twist." His primary message is that

> *The winning strategy does not just change as we move from stage to stage, it actually reverses the prior strategy.*
> —G. Moore [1995], p. 10.

Moore's work is especially interesting for the following insights:

- If product marketing strategies can change dramatically from one stage of the Technology Adoption Life Cycle to another, development strategies probably need to change also.

- Whether developing software for a software vendor or an internal IT group, one needs to understand what characteristics distinguish customers at each phase of the life cycle in order to successfully develop more appropriate products and thereby create greater customer intimacy.

- The Technology Adoption Life Cycle is a change model. It provides insight into topics such as learning and technology transfer.

The Technology Adoption Life Cycle

The Technology Adoption Life Cycle is a model of how people react to discontinuous innovation. It graphs as a bell-shaped curve that measures risk aversion, beginning with early innovators who are the first to embrace a new technology, and ending with the conservative laggards who grudgingly adopt a technology only when no alternative seems to be left.

Figure 7.1 depicts the Technology Adoption Life Cycle (shown as a bell curve of numbers of customers) divided into the kinds of customers in each segment.[1] The early market contains *technology enthusiasts* who are committed to and enjoy tinkering with new technology. They have little money, but they are essential to early marketing efforts—"they are the gatekeepers to the rest of the life cycle" (Moore95, p. 15). The enthusiasts are interested in the technology itself, not necessarily in its usefulness to the business.

---

[1] Figure 7.1 adapted, with permission, from Geoffrey Moore's *Inside the Tornado* (New York: HarperBusiness, 1995), p. 14.

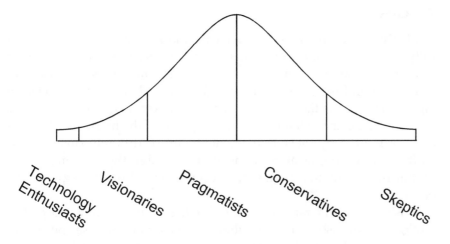

Figure 7.1: Technology Adoption Life Cycle Bell Curve.

The *visionaries* are early adopters of technology. "These are the true revolutionaries in business and government who want to use the discontinuity of any innovation to make a break with the past and start an entirely new future" (Moore, loc. cit.). They want to exploit new technology and make a mark, and they have money. Visionaries don't need recommendations from other successful users of the technology; they intend to set the stage. However, they do depend on the enthusiasts to bring new technology to their attention. There are a small number of visionaries in any market segment.

Pragmatists constitute the first high-volume segment of the life cycle. They form a major buying segment. Pragmatists are not looking for a breakthrough, but are more interested in a proven technology. They want to see it in use by other pragmatists, preferably in their own industry. Pragmatists want infrastructure and support in place before they commit their organizations. They want to buy from the market leader to insure reliability and support.

Conservatives enter late into the market. They are leery about the professed benefits of any new technology, and want solid proof of those benefits. They are also price-sensitive.

The last group, *skeptics*, are people who are so suspicious of anything new that they need to be sold *around* rather than sold *to*. Writers who are diehard typewriter users and are adamantly against computers and word processors are an example of people in this group.

The Chasm

While this life cycle model has been around for many years, it has not fully explained high-tech markets. Moore writes that the main reason it has not is the time chasm between the early-market enthusiasts and visionaries and the early-market pragmatists. This chasm, depicted in Fig. 7.2, represents the sometimes lengthy time period it takes to move from the visionary to the pragmatist segment of the market.[2] The buyers represented by these two segments are so different that a major shift in marketing strategy is needed to bridge the gap; moreover, within the pragmatist segment, a second marketing strategy change occurs.

Visionaries understand that a technology is new. They expect it to work, but they also understand they will have to improvise and probably provide substantial customization on their own. They are usually project-oriented, wanting to change only a relatively narrow area of the business.

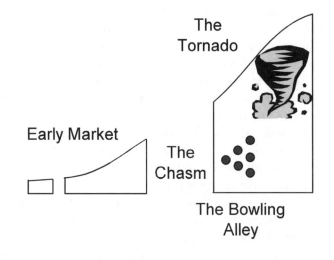

Figure 7.2: The Chasm.

Pragmatists want everything in place. They don't want to customize, and they expect support and services to be provided by either the primary vendor or specialized suppliers of services for the vendor's prod-

---

[2] Figure 7.2 adapted, with permission, from Moore, op. cit., p. 25.

ucts. They want to standardize the technology throughout the organization. They have a great deal of money. To sell to a pragmatist, a vendor must have good references and total problem-solution capability (software, hardware, training, service, and more), neither of which emerge from early market sales. This time chasm between the visionaries and pragmatists swallows many aspiring high-tech companies, particularly when forecasted sales to visionaries fail to materialize.

Moore's strategy for crossing the chasm and penetrating the pragmatist market is twofold. The initial effort invokes a niche strategy, which he calls "the bowling alley." Essentially, the strategy involves a company's carefully picking a market segment and working to provide a total problem solution for that *single* segment. Examples are Apple (with the Macintosh's penetration of the in-house graphic-artist market using desktop publishing) and Sun (with its targeting of the open systems market). When one segment is satisfied, the next is targeted (as if it were the next bowling pin in a bowling alley). As each segment is attacked, more of the *total solution* for which the pragmatists are looking is put into place.

When enough niche "pins" have fallen, the product enters another market phase. This is the point at which markets virtually explode as if a tornado were sweeping across the landscape. Some tornado-type successes have been Cisco Systems (with its explosion into a billion-dollar company in just a few years) or Netscape (with its virtually overnight capture of the Internet browser market). As the company enters the tornado phase of the market, the niche strategy is abandoned. The tornado is powerful and fast, sweeping competitors up in a swirling vortex. Once the pragmatists begin to buy, they move quickly. When the tornado is roaring, it sweeps up whichever product appears on the customer's doorstep. Oracle, for example, swept past the competition during the tornado phase, and has been a leader in the relational database market ever since.

Implications

I have spent more than a couple of paragraphs explaining Geoffrey Moore's work for two reasons. First, the volatile, high-tech markets he describes are exactly those that need adaptive development methods. Second, in order to adapt within the boundaries of a development project or a development organization, the members of that project or organization need to understand both the marketplace forces that drive their companies and the fact that those forces change quickly.

The most obvious effect of these volatile markets is the creation of a need for accelerated, adaptive methods to develop the products. A few months can mean the difference between being a market leader and a distant second. Moore discusses how the approach, or sometimes even hint, of a tornado product attracts huge influxes of money because the potential returns are so enormous.

> *These new pools of capital, in turn, create some of the fiercest economic competition on the planet, in part because winning or losing is compressed into such a short span of time.*
> —G. Moore [1995], p. 7.

The differences between the strategies in the bowling alley and in the tornado illustrate how an entire company must respond to drastically changing conditions. In the bowling alley, the niche strategy calls for high-contact, consultative selling. A premium is placed on understanding the customers' specific business and how they would apply the new technology. Product features are added for specific customers' needs. Customer intimacy is key to success. Since market segments are selected to avoid direct competition from larger firms, the focus is on the customer rather than on the competition.

Once out of the bowling alley and into the tornado, everything changes. There is only room for one market leader, and ascension to that throne will happen very quickly—especially noticeable in comparison to how slow movement was in the bowling alley. Moore describes the strategy as one in which the leader-to-be ruthlessly attacks the competition, expands quickly using fast sales cycles, focuses on operational effectiveness in order to deliver products, and ignores the customer. Ignores the customer? Once in the tornado, there is no time for customization or hand-holding. The strategy is to sell, deliver, and move on. Competitors who dwell too long on a single customer let four more get away to the competition.

There is an implication here for a product's mission profile and for its quality issues. Historically, a key selling point for methodology and process-improvement proponents was the reduction of total life-cycle cost by enabling developers to build the software right the first time, thereby reducing ongoing maintenance. However, in high-technology markets, notably during the tornado phase, the potential speed of revenue growth so far outstrips any long-term maintenance cost that the decision about whether to *do it perfectly* versus to *do it good enough* is really very simple. Miss the market, and long-term maintenance cost is the least of a company's worries. High-speed development strategies are critical in these environments.

A climbing analogy illustrates the point: The higher up a mountain one goes, the more quickly and more violently the environment changes. Recognizing when those changes approach and reacting to them quickly is necessary for a climber's success, and often for survival. Similarly, software development organizations that lock in to a particular set of strategies, or that do not adapt their strategies fast enough, cannot adequately support their companies. Understanding business and marketing strategies helps developers understand why they are asked to focus on one direction one week, and the next week on another. It's not because of arbitrariness or incompetence on the part of the marketing department, but because such behavior is necessary to survival in a turbulent ecosystem.

Product development requires periods of creativity and innovation, surrounded by intense rounds of practical implementation. It requires that developers have the flexibility to explore ideas hidden in unconventional niches and the doggedness to bring high polish and minimum defects to market. And, it often requires all of these diverse behaviors from the same individuals. It is what makes leading software product development in extreme environments such an exciting occupation.

Simplicity and Complexity

Good managers fail because they don't understand the difference between simplicity and complexity.

Command–Control management is insufficient to manage in complex ecosystems. Both silver bullets and Command–Control management seek easy solutions, craving a return to simplicity and simpler times.

I recently read a passage in an intriguing book called *Making Sense of Wine* that offered a view that, counter to popular opinion, we may actually prefer complexity over simplicity.

> *It appears that we are, in fact, set up to respond favorably to complexity. Decades of work in experimental psychology have revealed that when people are free to choose between a simple visual image and a more complex one, they gravitate to the complex. . . .*

> *What satisfies us so fundamentally about complexity is still the subject of speculation, largely in the academic field of aesthetics. It appears that we favor . . . uncertainty or lack of predictability.*
>
> *The more things are jumbled, the more 'information' can be conveyed at one time. . . . In short, there must be both pattern and uncertainty (complexity) for sustained interest.*
>
> —M. Kramer [1989], pp. 23, 24.

The final sentence of this passage crystallizes the philosophy behind Adaptive Software Development. ASD is a framework, a *pattern*, which provides some boundaries and some fundamental principles. At the same time, by being a framework, it remains open to, and moreover acknowledges explicitly, uncertainty and turbulence.

We are drawn to serial, non-iterative methods because they appear to be simple, silver-bullet approaches to managing complexity: "Just follow these 1,432 simple, sequential steps and you will be successful." But, possibly at some fundamental, philosophical level, what we really crave is a more holistic approach—one that speaks more to our sense of aesthetics than to our fixation on certainty.

Summary

➤ Disruptive technologies, and disruptive practices, can cause good managers to fail.

➤ Good managers fail when the rate of change disrupts their ability to cope using existing management practices.

➤ High speed is easy. The high change caused by high speed is much harder to manage.

➤ The ability to adapt to high-change environments can be a significant competitive advantage.

➤ In extreme environments, change becomes the norm and stability becomes the exception. While many people sense that the world is changing faster, most management practices still reflect linear mental models.

➤ At some level of change, Command–Control management based on predicted results falls apart. Beyond this breakpoint, self-organizing systems provide better solutions.

➢ Good managers fail when they attempt to use silver-bullet solutions to complex problems.

➢ There is no silver bullet, but there are Lone Rangers who have an arsenal of bullets for different situations.

➢ Silver bullets give us the *illusion* of stability, thereby reducing anxiety associated with the unknown. We use silver bullets to attempt to hide from reality.

➢ The optimization paradox shows that fads like TQM and BPR still rest on an underlying deterministic model. By trying to impose control on the uncontrollable, these programs are doomed to a downward spiral.

➢ The solution to the optimization paradox is to embrace an adaptive approach—one that recognizes and works with uncertainty.

➢ Good managers fail because their management practices were developed to solve complicated problems, but their problems are complex.

➢ Good managers fail because they don't understand the need for "requisite variety" in complex situations.

➢ Requisite variety is the antithesis of silver bullets. It states that managers must have a large number of coping mechanisms to combat the variety of changes in the environment.

➢ Good managers fail because they don't understand the business ecosystem in which their projects must live.

➢ Value discipline involves establishing a single business strategy for success: operational excellence, customer intimacy, or product leadership.

➢ Tornado marketing is about understanding the Technology Adoption Life Cycle and how companies have to change strategies quickly.

➢ Value discipline and tornado marketing help software developers understand extreme environments.

➢ The aesthetics of complexity are more important to us than the false security of silver bullets.

CHAPTER 8
Adaptive Management

"This need for control is so fundamental that it has dominated management literature for 100 years."
—A. De Geus [1997], p. 140.

Mountaineering in extreme conditions is not about skill, nor is it about strength and stamina—although they play an important part. Ultimately, extreme mountaineering is about judgment. It is about walking the narrow edge between success and oblivion.

Given good weather and stable ice conditions, thousands of climbers have the skill and stamina to climb Mount Everest, for example. Natural hazards—violent storms, avalanches, falling ice seracs, hidden cornices, precarious rock (which is waiting to be set loose by an errant foot or hand), high altitude (which can cause debilitating altitude sickness and death)—all present conditions that can change a climb from difficult to treacherous. Skill and stamina, understanding the risks of natural hazards, and good judgment in balancing rewards and risk (completing the climb versus the possibility of injury) are essential to making a successful ascent of big mountains like Everest, where the margin for error is very small.

How to manage the risk is not always obvious. For example, beginners often feel safest when tied to ropes, which are in turn attached to solid anchors. However, setting solid anchors, especially in variable snow and ice conditions, takes time. Taking time increases the risk that a climber will encounter bad weather, rockfall, icefall, and darkness. Time saps energy and concentration.

Speed is often the least risky course of action.

As the day wears on and the sun heats the mountain, the probability of rockfall and icefall increases dramatically. A climber crossing a long, dangerous, steep ice slope, on which a slip and fall would definitely be injurious if not fatal, must seriously weigh his or her overall ability in light of the possibility of getting taken out by a fast-moving missile from above.

During the descent off Mount Jefferson, two other climbers and I faced such a crossing. Not only was rockfall a *possibility*, but the glacier was strewn with rocks from earlier days and hours. The choice was not between which was a safe or a hazardous passage, but which was *least* hazardous. We opted for speed, albeit careful speed, pausing only briefly when a large rock thundered down between two of us who were descending about twenty-five feet apart!

Climbers call the leeway they have when making what could become life-threatening decisions the *margin of error*. Recreational climbers in moderate terrain have a wide margin of error. Their decisions could go either way, with minimal consequences. Occasionally, a decision made in moderate terrain results in a serious, but not devastating, outcome. Such episodes are called *learning experiences*. Enough learning experiences hone the judgment needed for more extreme climbs in which the margin of error is narrow.

There are several important points here that can be derived from thinking about climbing and applied to thinking about adaptive management: First is the point that speed is frequently the safest course of action. Based on my experience on dozens of adaptive projects, I know that customers are often so starved for results that they become ecstatic about whatever is delivered in three-to-six months. By the time a twelve-to-eighteen month or longer project is delivered, relationships between developers and customers may have been ruined, and reception of the product is unenthusiastic at best. With companies competing in the Internet software market, the landscape is littered with the slow and the bankrupt. Speed can also kill. On dangerous terrain, one misstep can send a person, or a company, plunging off the mountainside.

Second is the point that the terrain must match where *you* want to be. I, for example, will never climb Mount Everest. I am not good enough, either physically or mentally—Everest is terrain I choose not to tackle. Being partway up Everest is not the time to realize it is beyond my ability. Similarly, the middle of a project is not the time for you to realize your team is working on an extreme, complex project that it is not equipped to handle.

The third point is that the experience one needs in order to judge margins of error comes from testing one's limits in increasingly demanding environments. On treacherous terrain where the probability of a slip is high, keen judgment is a must. On moderate ground, one is safe to explore while judgment is learned. Beginners can get hurt on relatively easy terrain, but they usually are protected from disaster by a worthy emotion—fear. Intermediate climbers get in the most trouble; their technical skills often are superior to their judgment. High on the mountain, the climber's critical decision is always between advancing and retreating. Pushing limits is one thing, but ignoring risks is quite another. The best software teams and the best mountaineers are those who know when to advance and when to retreat. They understand the environment and its risks, and are able to balance their skills and the risks involved.

The fourth point is that decisions are made and actions are taken as the result of complex information and interactions. There are guidelines in the mountains, but few rules. Software teams that enter difficult terrain armed only with rules will fail. Rules can work on moderate terrain, but team members who know the exceptions to the rules hold the key to success on complex terrain. At the extremes, superb skill, honed instincts, and good judgment are the foundation for success. Skill and judgment allow the mountaineer to mitigate risk, but not eliminate it. Ignoring risk heightens one's dependence on luck; relying on luck alone is a poor long-term strategy.

The last point here is that mountaineers are great improvisers. While there may be a few rules that are rigorously enforced (for example, establishing a time to start down regardless of the team's progress), in the mountains "the map is never the territory." A forty-foot vertical cliff may nestle hidden between two contour lines on a map. Complex software projects need a goal, some guidelines, and possibly even a few rules, but success—much to the chagrin of rigorous-process proponents—most often results from effective improvisation.

Speed, proper terrain selection, intelligent risk evaluation, good judgment, improvisation, and an ability to understand the difference between rules and guidelines are all part of traversing dangerous software mountains.

The Adaptive (Leadership–Collaboration) Management Model

Is a new management model necessary? Does a model based on a better understanding of complex systems make sense? Is this a model we can act on today? One response to these questions came to me via e-mail from Tom Petzinger, a columnist for *The Wall Street Journal:* "Anyone who considers complexity actionable doesn't really get it. It's NOT a program. It's a way of thinking and sense-making." Petzinger, who regularly polls senior executives, believes this new way of making sense of the business world is the most important management trend of the decade.

Sense-making—that is, how we perceive the world around us—is indispensable to management. If we sense that the world is stable and predictable, our approach to management will be much different than if we sense that the world is turbulent and unpredictable. The sense that the world is relatively stable, a Newtonian view, led to management practices categorized as Command–Control. The sense that the world is turbulent and unpredictable is leading to a new set of management practices, sometimes labeled as participative, modern, or human-centered. In a turbulent world, one in which sense-making is enhanced by an understanding of complex adaptive systems, I think a more applicable term for these revised management practices is Leadership–Collaboration where "leadership" replaces "command" and "collaboration" replaces "control."

Commanders know the objective; leaders grasp the direction. Commanders dictate; leaders influence. Controllers demand; collaborators facilitate. Controllers micro-manage; collaborators encourage. Managers who embrace the Leadership–Collaboration model understand that their primary role is to set direction, to provide guidance, and to facilitate connecting people and teams.

Leadership and collaboration, the basis for adaptive management, are about creating an environment with the *requisite variety* to meet the challenge of extreme projects, particularly the challenge of high change. Iteration and concurrency required for high speed both create high levels of change. Without new techniques for containing these changes, high-speed projects of any size begin to rattle apart.

It is a generally accepted management principle that creativity and innovation are encouraged in relatively unstructured environments.

"[T]here is accumulating evidence that corporations fail because the prevailing thinking and language of management are too narrowly based on the prevailing thinking and language of economics. . . . [T]hey forget that their organizations' true nature is that of a community of humans."
—A. De Geus [1997], p. 3.

"The edge of chaos . . . is a state of paradox. . . . A group can be creative only if it holds the tension of conformity and individualism, which is only possible if the anxieties it raises are sufficiently contained."
—R. Stacey [1996], p. 152.

However, as size increases, the specter of imposed order creeps back into play. It is not that imposed order is wrong, but that it is insufficient. Visa International founder Dee Hock coined the word "chaordic" to describe organizations balanced on the edge between order and chaos. Others have used the term "poised." Adaptive management is about remaining poised on the edge of chaos and employing a variety of tools, some to create imposed order, some to encourage emergent order.

There are two factors to consider in the art of maintaining poise and creating emergent order as increasing size pushes organizations inexorably toward imposed order. The first consideration is cultural— one must create an organizational mindset that encourages and supports adaptation. Experience has taught me that it is not an easy task. The second consideration is structural—one needs to create a collaborative information structure that supports adaptive behavior across multiple and often virtual teams that are distributed in time or distance. Collaboration within a single feature team is enhanced by improving interpersonal skills. Collaboration across multiple teams is enhanced by creating an adaptive cultural environment and building effective structural support systems.

Adaptive Software Development involves a cultural shift from an optimizing to an adaptive mindset. This shift extends to management style, as was stated in Chapter 1.

Adaptation depends on Leadership and Collaboration rather than on Command and Control.

Adopting an adaptive management style can be difficult. One reason we revert to Command–Control management is that it has a history of working in large undertakings (although, as we shall see below, the history with large software projects is suspect). We are reluctant to apply the Leadership–Collaboration model to larger projects because there is a concern that it will cause chaos and failure. To encourage us to undertake large projects using the Leadership–Collaboration model, we need two things:

- evidence that the Leadership–Collaboration model works with extreme projects, and conversely, evidence that the Command–Control model doesn't work

- practices and tools to help manage volume (deliverables and changes, for example) as projects scale up, yet which at the same time support a self-organizing, adaptive philosophy

The Command–Control model has proven it can scale up on large projects—or has it? According to Capers Jones's figures, 48 percent of all projects with more than 10,000 function points as well as 65 percent of those with more than 100,000 function points are cancelled (Jones92). Many of Microsoft's products are in the 10,000-to-100,000 function-point range but far fewer than 50 percent of them have been failures. Why has Microsoft been so successful on large projects? The proponents of an optimizing approach to development (for example, the CMM) would argue that the failure rate on large projects is caused by lack of discipline. Microsoft is accused by these same individuals as being a bunch of undisciplined "hackers," yet its success rate is higher than the norm. In reality, producing a 50,000 function-point application by hacking it together is just not possible. Maybe Microsoft's success does not come from either optimized discipline or hacking, but from tapping into emergent order at the edge of chaos.

One source of this dilemma over whether or not optimizing approaches improve the success ratio on larger projects is the previously stated problem of confusing complicated and complex problems. The problem with viewing historical statistics is that many of those projects evolved in a more stable business climate than we are now experiencing. Few metrics are available for extreme high-change, high-speed projects. However, limited studies on RAD projects (those having up to 5,000 function points) have shown significant increases in speed and reduced costs as compared with traditional methods.

In addition to the success of Microsoft, there is impressive evidence that the Leadership–Collaboration model works for large undertakings, albeit outside the realm of software development. Witness a company that has grown 10,000 percent since 1970, continues to grow at 20 percent per year, operates in 200 countries worldwide, serves one-half billion customers, and has a sales volume of over $1 trillion—Visa International. Visa's success is attributable to Dee Hock's vision and his *chaordic* organizational philosophy that embraces adaptation and collaboration.

If the Leadership–Collaboration model is to support a chaordic organization, there must be methods and tools to provide order in key areas as size increases. As projects get larger and more complex and multiple feature teams become increasingly geographically and organizationally dispersed, more rigor is necessary. The quandary is how to support this additional rigor without stifling an adaptive culture.

One of the most thought-provoking questions about the application of complexity concepts to management is, "So, what is new?" Distributed decision-making, networked organizations, decentralization, organizational learning, distributed control—all have been management practices for some time, particularly in fast-moving industries, and should not be thought of as new. To answer what is new, I offer two responses. First, if the core of our belief system about managing organizations is rooted in the *old* science of deterministic Newtonian physics and survival-of-the-fittest Darwinian biology, then only a *new science* such as complex adaptive systems with an equally powerful philosophy and scientific foundation provides the credibility necessary for a major management cultural evolution. Second, the understanding of CAS does more than provide a powerful new conceptual base; it also helps us define, design, and develop new management practices and reinterpret and redesign existing ones.

Many companies apply some of the new management practices just listed, but the lack of a solid conceptual foundation restricts usage. Without a solid foundation, situational variations are more difficult to respond to, and new rules become as invariant as the old. Empowerment, for example, can be used as a critical element of a collaborative group, or it can be blindly pursued as an end in itself. How much empowerment or power-sharing is enough? Much of the management literature advocates one position or another (such as near-total empowerment through the use of self-managing teams), but offers little help in answering the question of how much power-sharing is enough.

An example of a CAS perspective on these questions would be Ralph Stacey's use of control parameters for balancing at the edge of chaos, as was described in Chapter 5. While complexity theory may not answer the question of power-sharing explicitly, it provides a different basis for exploring the issue. For example, one differing consideration for analyzing the issue of power-sharing would be whether the product market is orderly or complex. Since an orderly market tends to be managed using traditional methods, power-sharing is a difficult sell. In a complex market, managers implicitly understand that it is necessary to balance on the edge where power is shared and managers neither attempt to be in control nor abdicate control. The "edge of chaos" aspect of CAS, whether for power-sharing or for other practices, helps managers by providing an explicit rationale for some of their implicitly developed methods.

Second, the study of complex systems affords managers insights into new management methods and the redesign of existing ones. We have had hundreds of years to perfect hierarchical management, but fewer than twenty to deal with networked organizations. The movement in many of today's companies toward a flattened hierarchy (to deal with administrative issues) and networked teams (to actually produce products) is being performed with little data about how to *tune* those networks for effectiveness. The area of networked organizations is one in which a better understanding of complex adaptive systems may provide significant payback. How to tune collaborative networks is one subject of Chapter 10, and is, in fact, critical to scaling adaptive development.

In order to create the cultural and structural environment necessary to produce concrete results in turbulent, complex environments, we need to shift from Command–Control to Leadership–Collaboration management models.

Leadership

Success, according to author and leadership expert Warren Bennis, stems from three things: ideas—the intellectual capital of knowledge workers; relationships—people working together toward a common vision; and adventure—the willingness to take risk, to press on in the face of obstacles, and to choose action over inaction (Bennis89). Extreme projects require that all three of these attributes be finely honed. Great ideas cannot survive the lack of skills and experience. Without rich relationships in an environment tuned to exchanging and cultivating those ideas, the results will not be emergent. Without the leader's sense of adventure and the courage to risk making mistakes, the results will lack the spark that separates great products from good ones.

While the following fact may seem paradoxical, adaptive environments require much stronger leaders than do deterministic ones. The ability to help teams to understand the project's mission, to stand back and let the group struggle with mistakes, to encourage learning, to balance the need for flexibility and rigor, and to force decisions onto the

"[T]he key to competitive advantage in the nineties and beyond will be the capacity of leadership to create the social architecture capable of generating intellectual capital. And intellectual capital means ideas, know-how, innovation, brains, knowledge, and expertise."
—W. Bennis [1989], p. xii.

group all require greater leadership skill than drawing up a task list and commanding its execution.

Great leaders are people, too. Just like anyone else, they can have severe character flaws, and be capricious, vindictive, and petty. Even the venerable Disney "was an irritable, often small-minded, man . . ." (Bennis97, p. 201), but as a leader he was also an obsessive genius who was respected, if not revered. Some people may not like Bill Gates (even some at Microsoft), but few lack respect for his talent, his foresight, and his leadership abilities. Both men, Disney and Gates, demonstrate the difference between management and leadership—the difference is hard to define, but easy to observe. Success on complex projects requires good leadership.

> **Unfortunately, most software development projects are managed, not led.**

In order to battle complexity, leaders create collaborative environments in which solutions emerge and in which every member is accountable. Leaders must create environments in which ideas flourish, risks are taken, and mistakes are viewed as learning opportunities.

But it is not a Pollyanna world. Hard decisions must be made, products must be delivered, and consensus on decisions isn't always feasible. Ashby's Law of Requisite Variety requires the leader to know when to be flexible and when to be rigorous, when to seek consensus and when to dictate, when to exercise control and when to seek containment. Teams want a clear sense of direction and decisiveness from their leaders; they do not want arbitrariness or authoritarianism.

Leaders must be optimistic, pragmatic visionaries if they are to be effective. They make the mission come alive. They maintain a sense of balance and optimism in the face of unfavorable odds, but they also know when odds that are unfavorable evolve into impossible. Organizing groups of highly talented, opinionated, sometimes arrogant individuals can be like herding cats. They each want to veer off in an exciting new direction. It is the strength of the articulated vision as well as gentle nudging that brings the recalcitrant back into line, not authority. While the leader must be optimistic, his or her optimism should not be unbridled, but pragmatic. In particular, leaders who ignore risks are doomed to failure. Good managers acknowledge risks and manage them. They do not succumb to negativism.

Part of the practical visionary's job is to assess how the product conforms to the team's vision, to force hard product decisions, to know the difference between excellence and perfection, to select the best between different designs, and to know when to push and when to back off. The leader needs an intimate understanding of the product and the development process, but does not have to be the best developer. "Leaders of Great Groups inevitably have exquisite taste. . . . they are curators, whose job is not to make, but to choose. The ability to recognize excellence in others and their work may be the defining talent of leaders of Great Groups" (Bennis97, p. 200).

While pure visionaries are easily sidetracked (their minds are so full of options and opportunities that they cannot sustain the persistence and focus needed to ship products), practical visionaries ship products. A complex project needs a practical visionary.

Collaboration

Collaboration, as an activity in the basic Adaptive Development Life Cycle, focuses on interpersonal relationships within a project team. In the Leadership–Collaboration Management Model, collaboration entails building a structure in which large numbers of people can effectively interact.

A collaboration structure consists of two interwoven pieces: the component structure and the team network structure. In Chapter 3, components were defined as a collection of business functionality items or features that are planned and implemented together. The component structure defines the relationships among all the individual components being produced (for a more comprehensive discussion, see Chapter 9). This component structure helps managers and developers contend with the information flow arising from high-speed, concurrent activities that could easily lead to chaos. As was stated in Chapter 1:

> **The first major strategy requires deployment of methods and tools that apply increasing rigor to the results, that is, to the work*state* rather than to the workflow.**

The second collaboration structure piece, the team network structure, deals with ways of connecting people and teams. The team network structure defines a communications structure for each project (see Chapter 10). The interweaving of information and people creates an overall collaboration structure that is geared to producing emergent results—balancing information flows between chaos and stagnation. As we saw in Chapter 1:

> **The second key strategy requires deployment of methods and tools that support self-organizing principles across *virtual* teams.**

Within a collocated group, collaboration is concerned with the interpersonal values of trust, respect, and mutual participation. The structural components of collaboration defined within the context of the Leadership–Collaboration Management Model build on that base. Structural components are important in scaling collaboration to larger organizations, but the underlying interpersonal and cultural values are the most important.

Accountability

Ultimately, the success of any project rests on a team's ability to deliver results. And, delivering results must be the responsibility of every team member *and* every feature team member *and* every internal or external supplier to the project team.

Accountability is the process of ensuring that the right results get delivered to the right people at the right time. Establishing accountability and building collaboration should be complementary actions, but they are often antagonistic. The traditional practices for deterministic accountability stifle the creative juices needed for emergence. It is a dilemma—the working environment must be open and free-form, yet with definite boundaries (the results). Without accountability, there may be no product. But the traditional approach to accountability through increasingly detailed and precise optimization will not work.

Accountability appears to violate the creation of trust needed to make collaboration work. But in our imperfect world, trust eliminates neither misunderstanding nor infallibility. It is possible that someone may trust me to deliver something, but his or her expectations may be different from mine. Similarly, I may have every intention of delivering, but other priorities or simply momentary incompetence may delay delivery.

> **Accountability guards against fallibility, misunderstandings, and an occasional misplacement of integrity.**

Project management is in part about accountability. Time-boxed project management, for example, provides a framework for accountability while still encouraging adaptive behavior. In the end, if adaptive management fosters creation of innovative ideas but no products, it fails.

Creating an Adaptive Culture

"Organizational culture" has become a frequently used, and sometimes abused, term in recent years. Companies in the midst of marketplace upheaval try to change their culture to meet the challenge. At some level, culture is the sum total of belief systems, or mental models, that permeate an organization. Culture involves the explicit and the implicit—the "way it is" and "the way we wish it were." Many organizations, particularly those in high-technology markets, already have an adaptive culture, although few would identify it using these words. Making the mental models of an adaptive culture explicit may help such organizations to use the culture more effectively. Other organizations, particularly those with a culture built around a belief in imposed order, will have a difficult time making the switch. Two statements describe what is valued in an adaptive culture:

> *Adaptation* **is significantly more important than optimization.**

> **Emergence,** *arrival* **of the fittest, is significantly more important than** *survival* **of the fittest.**

Differentiating an adaptive culture from a traditional one are six characteristics: emergent order, simple principles, rich connections, distributed governance, poise, and balance. The first of these, "emergent order," was detailed in Chapter 2 and touched upon at various other points throughout this book. Belief in emergent order is core to adaptive development. Emergence arises from self-organization and is a means to generate results—not always predictable results, but results nonetheless. An adaptive culture encourages self-organization and emergence.

Next are "simple principles." Adaptive organizations work best when managed according to a few simple principles rather than with a myriad of simplistic rules, procedures, and methodologies. Managers who used some rules and methodologies—thought of as "cookbooks"—believed development could be reduced to a series of simple

"Pseudo-order is one maladaptive defense against uncertainty."
—J. McCarthy [1995], p. 99.

instructions and rote procedures. Adaptive managers, although they make best-practice material available to development teams, focus on a few guiding, clear principles that encourage complex, emergent behavior. A manager who says, "Let the system requirements evolve through iterative interaction with the customers" establishes a much different environment than one who points to a detailed procedural manual and says, "Do that." Innumerable simplistic rules encourage simple, uninspired behavior.

"Rich connections" are needed for collaboration, as was described in Chapter 5. Emergent results issue forth from the interactions and interrelationships embodied in a networked organization. Rich connections provide diverse information to a wide range of individuals working in concert. Principles and connections go together in an adaptive culture—simple principles, rich connections. Connections, collaboration, and networked organizations are topics of the next two chapters.

To create an adaptive culture, management must establish an environment in which distributed governance, poise, and balance also can endure. Since how this is done has not been fully discussed in other chapters, a fairly detailed treatment follows.

Distributed Governance

Adaptive organizations are decentralized, distributed, parallel, and even redundant, and they generally follow a policy of distributed decision-making. Team members' interactions are often confused, untidy, and illogical, but they work! An organization with distributed governance also has such qualities as empowering and participative management. Chapter 5 described what Ralph Stacey calls control parameters—characteristics of networks of people—which provide insight into how to better manage networked organizations. Distributed governance provides a way of managing what Stacey calls the degree of power differential.

My view on project team governance is that so-called self-directed teams, in fact, have no leadership. At times in the past, infatuation with *empowerment* may have led practitioners to assume that the need for leadership and leader decision-making no longer exists. Maybe there is a place on some projects for self-directed teams, but not on an extreme project. With all the emphasis given on extreme projects to collaboration, self-directed teams might seem justified, but that would be an erroneous conclusion. There is a big difference between a team led by a benevolent dictator and one on which all decision-making is abdicated to the group.

Whether we like it or not, power and politics are a fact of organizational life. In the fast lane, the optimal position is to share power. If Command–Control management anoints power through position, self-management seems to abdicate power. Leaders and teams empower each other. Leaders empower team members by assigning them components to develop and by making them accountable for the results. Accountability does not mean that a person should work in a vacuum. He or she must collaborate with other team members on decisions about the component, but he or she ultimately is responsible for the component.

Similarly, the team empowers the leader to make decisions. On a fast-moving project, there is not enough time to get consensus on everything. At first, the leader should consult with team members about decisions. As the team members' respect for and trust of each other and the leader mature, the leader can begin to make decisions without seeking team input (just as a developer can code a module without asking for input once the general design has been agreed to by the team). It is easy to make a mistake, however. Recently, having made a decision on a project and informed the group, I was interrupted by one individual who responded, "I don't disagree with the logic or the decision, but this was a decision you should have discussed with the group first." By that time, the reservoir of trust and respect was high enough for us to weather the mistake, but I felt properly chastised.

In extreme environments in which a wide range of information is critical to success, real power and decision-making authority are delegated from below, not above. Good leaders understand this. They also know they must make snap decisions occasionally, but they must do so cautiously and not arbitrarily.

A manager operating autonomously does not possess the requisite variety to manage an extreme project. Only through collaborative actions and joint decision-making can the myriad issues be addressed and changes be accommodated.

Software development projects are too big, too expensive, and too critical to an organization's success to allow a leader to abdicate responsibility. Good management is still important to successful projects, especially as project speed and complexity increase, but good leadership doesn't necessarily mean tight control of day-to-day activity. Although used in a positive way by Pascal Zachary in his book on the development of Windows NT (Zachary94), the term "under-managing" delivers the wrong message. Managing a collaborative environment is not under-managing so much as it is managing the environment and the culture rather than the details of how the product is produced.

Poise

Poise means to hold something, or even oneself, in equilibrium. For adaptive managers, one goal is to help both individuals and whole teams maintain their poise, to stay balanced at the point between inactivity and chaos where innovation and emergent results occur. In the context of the human psyche, being poised at the edge means staying in a state that is somewhere between the psychotic and the comatose. Keeping oneself and one's team poised on this brink, what Stacey calls "holding anxiety," and using the energy to generate innovation, is never easy. Balance deals with conflicting constraints. Whether a decision involves resources, customer requests, or which development practice or process to use, the judgment needed to maintain the proper margin of error is usually in short supply.

One problem leaders face is that although everyone understands the need for practical compromise, individuals usually have their mind oriented toward either rigor or flexibility. Once rigor-oriented individuals have established a goal and a path, they are virtual bulldogs in their focus. The flexible individuals cling to creative expression as a way of staving off bureaucracy. One mental model tends toward stability, the other toward chaos—neither balances on the edge. Therefore, one leadership task is to help individuals of either mindset maintain the poise required to balance the needs for rigor and flexibility.

A subset of the need for poise is the need to focus and de-focus. In the midst of seeming chaos, when others are racing in ever-widening circles, a leader's ability to concentrate on the mission and to focus his or her team on the desired results is critical to success. Time-boxing, for example, is a technique to create just such a singular focal point. In a complex environment, there needs to be a simple focusing variable, but it cannot be inflexible. The ability to de-focus, to recognize a new situation and adapt to it, is equally important.

In basketball, teams tend to either run set plays or execute motion offenses. Set plays are run regardless of what formations the defense establishes. Motion plays adapt depending on defensive response. An Adaptive Development Life Cycle is like a motion offense, adapting to overcome the obstacles as they arise—but in order to sustain success, good leadership and judgment are requisite. A team short on leadership and judgment should stick to set plays.

Poise is about both understanding and using paradox—building teams while honoring individual identities; implementing adaptive practices while understanding the need for some rigor; focusing on the goal while leaving the mind free to speculate; and changing continuously while always maintaining a stable base. Leaders who can use paradox and contain anxiety—both their own and their team's—and who know when to compromise and how to manage trade-offs are the survivors who will succeed in managing complex, extreme environments.

Compromise

Compromise means settling differences through each side making one or more concessions. A trade-off involves exchanging one thing for another, the relinquishing of one benefit for a more desirable or more acceptable one. Unfortunately, both compromise and trade-off are words tinged with moral overtones, but they describe necessary behavior. Most development staffs understand the need for trade-offs. They might not like to cut features to meet a schedule, but they understand. One trade-off technique, quality function deployment (QFD), centers on the creation of a matrix that arrays all quality characteristics and correlates whether pairs support or conflict with each other. For example, reliability and maintainability may be mutually supportive characteristics, whereas ease of use and functionality might be conflicting characteristics. Software engineers, hopefully with their customers' input, make these trade-offs constantly. These are compromises on things, which, while difficult, do not usually impact a team member's value system. Compromising on values and beliefs is much stickier.

The word "compromise" has a shabby connotation. The negative image is attached to the idea of a person's conceding important beliefs or values. Compromise is usually thought of as occurring between two or more people, but in reality, we compromise within ourselves every day. Every decision requires a juggling of information and mental models—a compromise. For example, most of us believe in the moral principle that it is wrong to kill, but we see exceptions to the principle nearly every day.

- It is okay for society to kill (capital punishment).

- It is okay to kill in self-defense.

- It is okay to kill as an act of war.

- It is okay to simulate killing (such as is shown on TV, in movies, and in video games).

- It is okay to kill animals for food and clothing.

There are also purely case-specific exceptions. For example, someone may have a moral objection to capital punishment—except when someone close to him or her was the victim. All mental models have flaws. Our own mental-model combinations may run the gamut from supportive to conflictual. Even though we seldom make it explicit, we each have a value-correlation matrix in our mind similar to a QFD matrix. Every time we make a decision, we access this admittedly fuzzy matrix to weigh possibilities and make our own trade-offs or compromises. Based on personal experience, I can state that group collaboration never occurs without compromise.

So, on one hand, we consider compromise to be an affront to our values and therefore distasteful, but on the other hand, we know compromise is absolutely necessary to attain our goal. Part of this problem with the word "compromise" may be that we need finer distinctions than a single word offers. The leader of an adaptive group must learn three types of compromise—synergy, mutual concession, and appeasement—but avoid a fourth—tyranny.

Synergy means combining operations or actions. In adaptive groups, it means getting something without giving up anything. A synergistic outcome is one in which the sum of one plus one adds up to significantly more than two. Everyone's values or mental models are accommodated in the outcome. The mental models are supportive, not conflictual. One measure of how much a team has jelled is its degree of synergism. However, striving for synergy—that is, for complete agreement or mutually supportive outcomes—every time is unrealistic, especially in high-speed environments. Even if synergy occurs only 20 percent of the time, that may be enough to create a very powerful team. Synergy makes a tremendous contribution to collaboration. For example, suppose that I believe in code reviews, but you don't.

However, in our discussions, we agree on several situations for which we *both* think reviews will be beneficial to creating a better product—the compromise comes as a combination, or synergy, of my ideas and yours.

Mutual concession means giving up one thing for something else. In adaptive groups, the goal of mutual concession is to create something of greater value than what was given up. For example, I may be willing to give up some part of my autonomy to be part of a team community. I may decide, after appropriate discussion, that since the team as a whole thinks code reviews are a waste of time, I will forgo their use on this project. What I may get back is an enhanced feeling of being part of a team. By giving a little, I have contributed to the jelling of the team. Sometimes, I concede something; at other times, someone else does. In this way, mutual concession plays a role in the majority of decisions during even a jelled team's life span. Compromise is not a necessary evil—it is just necessary.

Appeasement means giving up one thing for something less. Appeasement becomes the search for the lowest common denominator. For example, I give up code reviews, but do not get back enough to feel very good about my sacrifice. Whereas synergy and mutual concession contribute positive energy to group collaboration, appeasement absorbs energy. Appeasement may be necessary in a few situations during any project, but project health deteriorates with overuse.

Tyranny means being forced to give up something for nothing. With synergy, mutual concession, and even appeasement, each team member has a degree of choice, or at least an opportunity to contribute significant input. Tyranny occurs when decisions are imposed through the blatant use of power without regard to any input. Making trade-offs through the use of tyranny instantly kills collaboration.

The leader's job is to understand these various views of compromise in order to balance each individual's self-interest with the team's mission.

Managing the Emotional Roller Coaster

Adaptive projects are emotional roller coasters. Adaptive management involves leading people to the edge of chaos and keeping them there for a prolonged period. As the project progresses, waves of chaos and accompanying strong emotions roll over the project team. Virtually from day to day, team members can vacillate from euphoria to near depression. The project manager who stands in front of a wave and tries to fix everything and make it better will get washed away. Project managers accustomed to more traditional approaches have a lot of unlearning and relearning to do.

A critical challenge for the project manager comes from team members asking for *fixes*, either directly or indirectly. The manager must respond not only to the team members' emotions on the roller coaster but often must also fend off his or her own emotional swings. The entire team needs to understand that these emotional swings are normal and not overreact. Knowing that emotions are part of the process does not reduce the emotional level, but it is comforting to understand that they are normal.

But don't underreact either! Adaptive development is frustrating because it is a balancing act. Emotions are an important component of project feedback. The question to ask is, "When are they normal reactions to the ambiguity and pressure of the situation, and when do they foreshadow problems needing attention?"

Where does all this emotion come from? First, although the team has developed a mission, including an outline of the deliverables, there are still many *unknowns* about the final product. Especially during the first cycle or two, team members can feel insecure about the project's outcome. Insecurity, combined with the pressure of short-term deliverable deadlines, leads to anxiety and stress. On one day, the feeling may be of exhilaration at something suddenly falling into place, and on the next day, hopelessness about whether the deadline can ever be achieved.

The learning process itself is emotional, particularly the learning that takes place while a team is delivering partially completed products and then listening to customer feedback. There is heightened anxiety about whether the customers will like the products. There is anxiety because the team may not be used to releasing partially completed products with known flaws. There is anxiety about the damage to one's ego, about one's reputation in being associated with mistakes and partial results. I call this the Colombo School of Development. Colombo got much more information from playing dumb than from playing smart, but playing dumb is a persona most people find hard to adopt.

Another trigger for strong emotion occurs when development techniques are brought in to bolster the sagging fortunes of groups perceived as doing a poor job for their customers. Customers are already unhappy, and the new techniques are intended to change those perceptions, but the developers see themselves facing a hostile audience that may be ready to pounce on their first mistake.

It might seem that these emotions are present in more traditional (Monumental) projects, but they are not often of the same intensity and frequency. In a traditional, serial project, the team usually is insulated from customers by documents the customers don't understand. At the end, developers are so worn down from months and months of work (with little real feedback) that the only real emotion is relief that the project is over. Since there is no way the product can be reworked at this point, customer criticism is viewed as irrelevant.

The bottom line is that traditional projects are less emotionally charged, with fewer intense swings, than adaptive, short-cycle projects. For staffs accustomed to benign emotional environments, extreme projects can be a rude awakening. Just as getting to the top of a mountain peak involves a climber's experiencing fear, boredom, and exhilaration, implementation of adaptive techniques involves dealing with powerful and frequent emotional swings.

Teams practicing adaptive development are poised above the abyss. Projects with this profile are both high-reward and high-risk. Teams must be incredibly focused, but can't be dogmatic. Too much dogmatic adherence can bring stagnation. Too little discipline, too much hot-dogging, brings chaos. Like a mountaineering adventure that tests the limits of human capability, high-speed, high-change projects place extraordinary demands on the physical, mental, and emotional well-being of the development team. But they also heighten the sense of reward and accomplishment.

Holding Anxiety

A leader of an extreme project must be able to distinguish between high anxiety and the onset of serious dysfunction in order to help the group deal with constant pressure. The edge of chaos is an uncomfortable place and some group members will push for more rigorous processes to reduce the terror of the unknown. Once, at one of the most cutting-edge high-tech software companies I've consulted with, I had a developer suggest that the team *freeze* requirements early so it could get the product out the door—clearly the request was a plea for reducing anxiety. Other teams under pressure will push for *less* rigor—

so they can retreat into the less stressful situation of doing whatever comes to mind.

Good leaders know how to help the group *hold anxiety* at the balance point at which innovation and drive will be sparked, without ultimately frustrating everyone. One important task for the leader, especially in groups new to high-speed environments, is to help team members understand that the feelings of anxiety are natural, that everyone has them, and that they are not going away. Knowing that everyone is feeling the same, even the leader (in fact, particularly the leader), helps people. On real projects, where the stress is high and learning requires an admission of fallibility, emotions can be explosive and leaders need an arsenal of techniques, as well as perceptiveness, to help the team hold anxiety.

Good leaders also help their group relieve tension. Whether it's by encouraging a fast-paced ping-pong match at lunch, or by making someone go home early, these leaders have good intuition when it comes to detecting stress levels. In the software business, the measure of a truly great leader is his or her ability to keep the anxiety at a level where team members want to return for the next project. A situation in which everyone jumps ship at the end of a project indicates that the group got pushed over the edge, but managed to hold on briefly—this does not equate to a long-term strategy.

Accidental Success

"Leaders of Great Groups trade the illusion of control that micromanaging gives for the higher satisfaction of orchestrating extraordinary achievement."
—W. Bennis and P. Biederman [1997], p. 214.

During an adaptive project I consulted on several years ago, the developers were dismayed at their work plan. This four-month project had customer focus-group reviews scheduled once a month and, at the conclusion of the initial JAD and project-planning sessions, we told the development team to have the first-cycle features ready for the focus-group review at the end of the month. Accustomed to more traditional task plans, the developers' first question was, "What do you want us to do next week?" My response, "Whatever you think it takes!" nearly led to mutiny.

By specifying only the results desired and the time frame in which to attain them, I focused team members on being responsible for the "how"; they became the owners of the requisite variety. One team member remained unconvinced this approach would work, even after successful on-time delivery of the final product. Lacking a definitive task plan—a defined path from starting-point A to result B—he viewed the success as accidental.

One of the most difficult emotional barriers for adaptive managers is the gnawing feeling that success is accidental. Without clear cause and effect or clear commands, project members and managers alike sometimes see the results as accidental. A manager cannot always point to *why* something succeeded. In fact, because of the anxiety, the chaos, the need to under-manage and push decision-making to the lowest level, leaders often feel *more* frustration than do group members. When, in addition to all this, *mistakes* must be accepted as an integral part of the process, becoming an adaptive manager starts to be a very difficult emotional undertaking.

Balance

Whereas poise is about people, balance is about things, maintaining trade-offs between product characteristics and the practices used to create those products. The mission profile in Chapter 3 provided a tool for balancing product characteristics. Balancing rigorous and flexible practices (or processes) is another important leadership activity. Some parts of the development effort are more certain and can, therefore, be managed with optimizing practices such as configuration control. Other aspects of development are more uncertain and unstable and need to be managed with adaptive practices such as speculating (rather than planning). The ability to properly judge the situation and to decide what tools to use are also necessary skills if one is to be an effective leader.

Our ability to balance is clouded by traditional definitions of process. The next section examines *process* from the perspective of stable, unstable, and complex environments and then examines why using patterns, rather than process, is advisable in complex environments.

The Progression from Process to Pattern

Extreme projects seem chaotic. Everything appears to be unstable—business practices, technology, and competition all explode with changes driven by increasing complexity. Managers use one of three basic strategies to cope with increasing uncertainty:

- They retreat from complexity, by refusing either to change or to adapt (a poor strategy for long-term survival).

- They try to control complexity; but as complexity increases, they must try harder and harder for diminishing results.

- They adapt to the complexity, by trying to bound or contain changes and by maintaining the flexibility to react to any event.

Whether a manager attempts to control or adapt to complexity, he or she uses certain processes or practices. Unfortunately, process improvement initiatives such as the CMM and Business Process Reengineering have imposed on the word "process" a connotation that suggests strict definition of inputs, results, and process steps—as if even creative activities could be "engineered." But, as we broke down meanings of the word "compromise" in the preceding section of this chapter, we should similarly dissect the word "process" in order to use it more effectively.

One necessary distinction stems from software development paradigms of the past, in which developers and managers oscillated back and forth between *no* process (Accidental Software Development) and *burdensome* process (Monumental Software Development). Such efforts aren't effective today. The emerging solution is to neither abandon process nor force-feed process, but to *steer* process—adapting methods to fit the situation using careful analysis and relevant feedback measures. One of the reasons I use the analogy between software development and mountaineering is that both are *steered* activities. Venturing into the mountains without the proper skill set is a sure recipe for disaster. Having a mental attitude that allows no variation in the face of changing environmental conditions is even worse. This is the case in software development, too.

One debate in software development circles is whether the most important component is *process* or *people*. As you may have supposed, one camp emphasizes process as more important than people, while the other stresses people as more important. It's not that process-oriented advocates regard people as unimportant; they just regard them as secondary. Adaptive development is built on the principle that people *are* the most important ingredient for success, and that process should play a supporting role. While processes (and, as we will see, patterns) are necessary, they must be implemented in such a way that they support project teams as organic, living ecosystems.

A Process Classification

Processes are organizational structures that focus human energy on producing results.

The important words in this definition are "organizational structures." In knowledge work, the environment encourages self-organization, which in turn generates innovation and emergent results. From W. Edwards Deming to more recent process-improvement writers such as Geary Rummler and Alan Brache (Rummler90), the emphasis is on improving the environment—creating clear tasks, assuring minimum interference, and defining consequences for actions. Organizational structures guide work and affect the product. Structures can encourage innovation or destroy it. They can create boundaries within which teams can produce phenomenal results, or which stifle every ounce of vitality a team possesses.

Organizational structures influence not only product features, but also the evolution of personal relationships. For example, the Waterfall Life Cycle was developed to help *control* software development through the successive production of specific deliverables. This waterfall organizational structure tends to create a work environment in which detailed, step-by-step processes are valued and personal relationships gravitate toward power and control. Iterative or Adaptive Development Life Cycles emphasize flexible processes and collaborative personal relationships among developers.

There are process-improvement practitioners whose underlying philosophy seems to be, "If we improve the process, better results will automatically occur." For example, the SEI's Capability Maturity Model measures processes in place, not results. An organization's "maturity" is assessed by checking off a list of required processes for each level—inspections, project task-scheduling, or unit-testing, for example. From a CMM perspective, level 5 is better than level 4—more processes mean higher maturity. However, there does not seem to be an evaluation of whether a specific level is appropriate for a particular type of product development or whether companies at level 5 produce better results (according to customers) than companies at level 3 or 4.

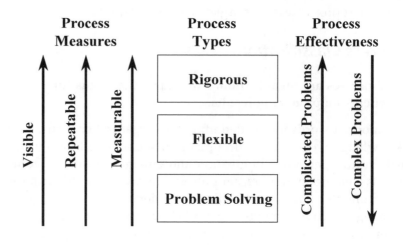

Figure 8.1: Process Classification.

A process classification scheme is shown in Fig. 8.1. The characteristics along the left side of the diagram—visible, repeatable, and measurable—describe attributes of a process. Processes that become more visible, more repeatable, or more measurable are *better* (as depicted by the *process effectiveness* arrow on the diagram), according to traditional process-improvement proponents such as the SEI. As processes become more rigorous—that is, as they move up from problem-solving to flexible to rigorous—they bring efficiency and lower cost. They *optimize.* For complicated problems in orderly environments, increasing rigor usually increases process effectiveness.

But what about the processes' ability to handle complexity and disorder? Disorder arises from differences in initial conditions, from wide swings in inputs, and from having ill-defined input-to-output conversions. Disorder results from uncertainty and unpredicted changes. Rigorous processes actually have a tendency to ignore inputs outside some predetermined range. For example, developers may ignore an additional customer requirement because it is outside the defined scope of the project. While limiting scope "creep" is an important project management tool, it can blind the team to inputs that indicate the scope itself might be incorrect. More rigorous processes are effective in managing within the stable portions of the environment. They are insufficient in the unstable ones.

The more universal a process, the more flexible it needs to be.

A corollary to this observation is,

The more rigorous a process, the narrower its applicability.

Running an entire project with only rigorous processes will lead to failure if the environment is uncertain, disorderly, and unpredictable. What about running a project with entirely problem-solving processes? The problem with this is the failure to take advantage of dealing with the orderly parts of the environment. In a project of any size, the disorderly portion is hard enough to deal with, but if poor management causes the potentially orderly part to become disorderly, problems are magnified. Dynamic systems amplify feedback. For certain disturbances, such as a low level of defects in a system, the impact of the next defect is linear. But, at some inflection point, the impact of the next defect becomes non-linear, rapidly pushing a project into complete chaos.

A simple example would be a project with three processes: requirements specification, code development and test, and configuration control. For example, imagine a project Zeus run as a problem-solving effort. The developers wait until a problem occurs and then fix it—they use no requirements documentation, no coding guidelines, no configuration control procedures. For small projects, problem levels may be low enough for this approach to work. But as project size grows, problems begin to multiply and progress slows. It may be that for this particular type of project, it is nearly impossible to reduce the number of requirements changes, but the changes are compounded by the fact that code and test-case changes are not controlled and the project becomes unstable. If we ask, "What are the most predictable problems on this project?" the answer will be, *configuration problems*. Putting a more rigorous process in place for configuration problems may be all this project needs to put it back into balance.

An adaptive project manager knows when to use rigorous processes and when to use more flexible or problem-solving ones. He or she also knows that as uncertainty increases, the usefulness of rigorous processes will decrease rapidly.

Rigorous Processes

Rigorous processes deal well with potentially orderly parts of a project. To be *potentially orderly* means that something is predictable; therefore, its predictability can be used to advantage. Teams not taking advantage of this predictability miss a source of stability in complex environments. Rigorous processes are repeatable, visible, and measurable. For a process to be visible, it needs to be well-defined. That is, the process must be written down and understood by the people who have to execute it. To be visible also means that the conversion of inputs to outputs is a well-defined, even algorithmic, conversion. Rigorous processes can usually be refined to achieve high levels of productivity and can also, therefore, be considered optimizing in nature.

A range of acceptable input conditions should be defined for a rigorous process. Without some idea of these acceptable conditions, the executors of rigorous processes may try to apply the process where it is inappropriate.

Typically, rigorous processes are boring—almost administrative in nature. Since they are less fun than more flexible processes, rigorous processes require discipline, on the part of both team members and management. Poor execution of rigorous processes can easily tip a mildly unstable project into complete chaos.

Flexible Processes

Many of the processes used in software product development are flexible rather than rigorous processes. Flexible processes can accommodate a wide range of inputs, the steps in the process may be fuzzily defined, and the results are not as repeatable when compared to those of rigorous processes. Rigor often is most applicable to process *documentation,* while flexibility is applicable to the analysis of what goes into the document. For example, the essence of any model-driven development is not the drawing of models, but the analytical thinking effort leading to the drawings.

Flexible processes are less precise than rigorous ones. For example, they may not be defined formally, they utilize guides rather than rigid rules, or they may be applied differently by different teams in the same company or even within the same product group. A flexible process may not produce the *same* results every time, but the results are *similar* enough that some written description of the process will benefit the organization.

Problem-Solving Processes

Problem-solving processes should be used where inputs are unpredictable or process steps ill-defined. For example, we could establish a general process for responding to a competitor's actions, but the process might be so general as to be useless. We know that competitors might introduce a new or improved product, but we do not know in advance what the new features or price might be. In this case, the project plan says, "If competitors change features or pricing, we'll deal with the new situation as a problem when it occurs."

Problem-solving is at the creative heart of software development. Where a rigorous, repeatable process is applicable, the result is seldom in doubt—execution of the steps produces the desired result. Conversely, a problem-solving process is far from a sure thing—the team may not find a solution, or the one they do find may be wrong. So, while we can surround problem-solving with more rigorous data-gathering or documentation processes, it is important that we remember that problem-solving is an emergent process—one that defies strict cause-and-effect analysis.

Patterns

Process classifications are still insufficient to understand producing results in complex environments. Because of its history, the word "process" still sounds too mechanistic. We need to move beyond the process reengineering community's view that flexible processes are those just waiting to be "fixed."

The fields of cybernetics and engineering feedback systems have driven our view of process and process improvement. Process management books abound with figures similar to that of Fig. 8.2, which depicts little machine cogs, reacting to well-known inputs, grinding through a set of predefined activities, and producing results within well-established boundaries. Measurements of the results are compared with plans, and deviations are controlled by appropriate actions. Even those who admit to the necessity for process flexibility seem to hold this model as their primary vision. For example, the controller is typically compared to a thermostat in a heating system, monotonously adjusting for deviation from the goal.

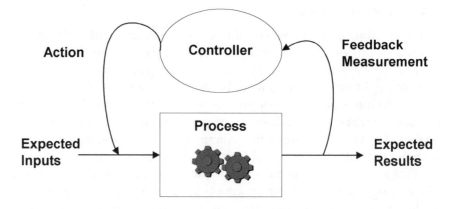

Figure 8.2: A Mechanistic System Diagram.

Figure 8.3 depicts a very different vision—one based on organic rather than mechanistic principles. Patterns replace process—patterns formed by the shared mental models of the individual people who are trying to forge the results. Many of the differences between Figs. 8.2 and 8.3 embody what we have learned about complex adaptive systems.

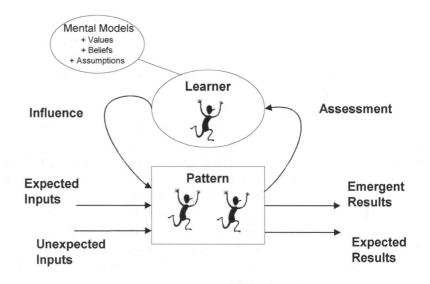

Figure 8.3: An Organic System Diagram.

First, in complex ecosystems, we expect the unexpected. Rigid processes respond to the unexpected by throwing the input away; if it is not expected, then it must be extraneous. Patterns welcome the unexpected as a source for learning. Second, there are people, not cogs, inside the box. People can do similar things repeatedly, but never the same thing. With a step-by-step process we expect the same set of outputs from identical inputs, but a person's reaction to the inputs may vary considerably from day to day based on a wide variety of conditions, many of them unrelated to the task at hand.

Third, while patterns may produce expected results from expected inputs, they also produce emergent, or unexpected, results. This is one advantage of having people rather than cogs in the middle.

Fourth, to the Command–Control manager, governance means measurement. The notion of control, particularly in a highly process-driven organization, involves increasingly precise measurement, comparison with nominal values, and decisive action. This idea of feedback measurement—in most cases, meaning evaluating people—is firmly ingrained in our traditional management culture. Robert Austin (Austin96), however, provides a very different view of measurement.

"mea·sure·ment dys·func·tion n : compliance with the letter of a motivational scheme in such a way as to achieve exactly the opposite of that scheme's underlying goals and intentions"
—T. DeMarco [1995], p. 24.

Austin's organizational performance model states why so many measurement programs fail. Based on economic theory, Austin's model builds a clear picture of the difficulties in motivating through measuring, particularly in knowledge work. He defines measurement dysfunction as measuring something to get a particular result but in which the act of measurement causes exactly the opposite result to occur. As Austin shows, reliance on simple measurements in complex situations nearly always leads to dysfunction.

Process improvement initiatives strive to create repeatable processes. To be repeatable, the processes must be measured, but as we all know, process measurements are in actuality people-performance measurements. Therefore, if Austin's theory is correct, process improvement initiatives have a reasonable probability of generating dysfunctional behavior over time. Patterns use general, multidimensional assessments rather than reams of strict metrics. Reasoning from Austin's theories, one can see that the use of patterns should result in less dysfunctional behavior.

Finally, an organic system gathers a wide variety of inputs—those that can be measured and those that cannot. An organic system might assess both the number of defects found per thousand lines of code and the emotional distress of the frazzled test team. An organic system is not as concerned about comparing actual versus planned and taking some "control" action as it is concerned about learning whether the plan was any good in the first place. An organic system cares less about controlling how people behave than it cares about influencing them to consider alternative courses of action.

An adaptive manager views the world as organic rather than mechanical. The view is yet another element in creating the different cultural perspective required to produce results in complex environs.

Poised at the Edge of Chaos

Adaptive management is formidable because it requires multiple behaviors (leading and collaborating) from the same individual. Walt Disney was a model for adaptive leadership. Disney's original animated film, *Snow White and the Seven Dwarfs*, was 83 minutes long, required 250,000 finished drawings, and included the work of hundreds of artists (Bennis97). Filming demanded artistry, obsession, pursuit of excellence, and attention to detail—a delicate balance between artistic expression and rigorous assembly and production. Walt Disney was a leader who knew how to balance these conflicting behaviors in order to deliver a film now considered to be a classic.

Developing sufficient judgment to balance rigor and flexibility may be the most difficult adaptive skill to learn. Too little rigor can easily throw a project into chaos, where confusion, rework, and short tempers create an accelerating downward spiral from which recovery is unlikely. Too little flexibility saps creativity.

Mountains generate patterns and rhythms—climbers respond. Weather patterns of wind and snow; slow glacial advance interrupted by widening crevasses or toppling ice towers; daily cycles of freezing and thawing; patterns of cracks, nubbins, slopes, and indentations in the rock—these are all conditions that climbers must interpret and respond to accordingly. Each climber and each climbing team responds differently to the challenge of a particular rock face, just as software developers and teams respond differently to a particular development challenge. Competitors are no more predictable than mountain weather.

There is no Command–Control strategy in the mountains—no competent mountaineer concedes responsibility for his or her life to someone else. But, in a climbing team of passionate, often egotistical individuals, every person also understands the need for both leadership and the entire group's involvement in critical decisions. A Leadership–Collaboration model is as appropriate for a software development team as it is for a climbing team. The skills required are similar—superb technical skills, razor-sharp judgment, a knack for improvisation. The attitudes required are similar also—respect for the environment and an understanding that it, not you, is in control. With extreme projects, whether the goal is a mountain summit or a software product, continuous adaptation is the only strategy that works.

Summary

> Speed is often the least risky alternative. The margin of error hangs on the right balance between speed and environmental hazards.

> Success in extreme environments requires a combination of skills and judgment—and luck. Judgment, by far, is the most difficult skill to acquire.

> Adaptation depends on Leadership–Collaboration rather than on Command–Control.

> Collaboration in the management model focuses on deploying methods and tools that support self-organizing principles across *virtual* teams.

> A key structural component of adaptive management involves deploying methods and tools that apply increasing rigor to the results, utilizing the work*state* rather than the workflow.

> Accountability guards against fallibility, misunderstandings, and an occasional misplacement of integrity.

> The characteristics of an adaptive culture are

  • emergent order

  • simple principles

  • rich connections

  • distributed governance

  • poise

  • balance

> Leadership in adaptive projects is about setting directions, creating environments, and letting results happen. Under-managing is a difficult task for many managers to achieve.

> Empowerment goes both ways—from leader to team member and from team member to leader.

> One of a leader's most difficult challenges is helping the group maintain *poise* and *hold anxiety*.

> "Process" needs flexibility. What is helpful is a process classification in which appropriately rigorous, flexible, or problem-solving processes are applied as required.

> The word "process" is too mechanical. "Patterns" provide a better way to manage organic, people-driven systems.

CHAPTER 9
Work*state*
Life Cycle
Management

O ne of the primary goals of this book is to provide a framework for organizations that need an adaptive approach on large, complex projects. Scaling up from a model is a problem in many fields, but it is especially critical for engineers. I once worked for an oil company that had a project to construct a fifty-million-dollar pilot plant that was intended to handle a new high-temperature, high-pressure oil-refining process. On a small scale, the refining process worked fine in the laboratory, but the company was leery about spending a billion dollars to build a commercial facility without first testing the process by means of a realistic, working model. Scaling up from the lab model to a commercial plant was too big a risk, even though all the engineering studies appeared sound.

The test plant blew up, scattering the foot-thick walls of the containment tank over the landscape. The commercial plant was never built. The fifty-million-dollar pilot saved the company from a billion-dollar mistake. Under day-to-day operational conditions, the extremes of temperature and pressure were just too unstable. Scaling is always an issue in engineering. In software engineering, we often forget that the problem differences between a thousand, a ten thousand, and a hundred thousand function-point project are not linear, but exponential.

"Because product development hinges on problem solving (which is notoriously unpredictable), strict automation of everyday tasks can destroy the creative energy which drives top developers."
—W. Collier, D.H. Brown & Associates, private correspondence [1996].

Our traditional approach to managing larger projects has been to increase the rigor applied to the development efforts. Tools are oriented toward stabilizing the uncertainty and change by employing project-management, configuration-control, requirements-specification, and other techniques. These techniques seemed to work well through the 1970's and most of the 1980's for projects undertaken in large banks, insurance companies, telecommunications firms, and in the military. But while many of these projects were *complicated,* they were not *complex* in regard to their need for speed and for rapid rates of change. More rigor became the approach used to handle increasingly greater size and complexity. Because no distinction was made between *complicated* and *complex,* increased rigor was applied as the answer to every problem.

But as the forces of globalization, mass customization, extended enterprises, and electronic commerce led to greater complexity, our rigorous tool set began to fail us. As if with blinders on, we still fail to learn lessons from software developers such as Microsoft, whose Windows NT, reportedly containing upwards of 35 million lines of code, is not only huge—which, in its own way, contributes to complexity—but it must survive in a highly competitive environment in which speed and change are the dominant forces.

Large, complex projects require increased rigor, but they also require an adaptive, on-the-edge, collaborative environment in which to generate emergent results. These projects must be simultaneously rigorous and emergent, and they pose a very difficult leadership challenge.

There are three solutions to the challenges of large, complex projects:

- An adaptive culture, as described in Chapter 8, is needed.

- Rigor must be viewed as a balancing force, not as an end goal.

- Frameworks must encourage emergence at the same time as they bring increased rigor to certain practices.

Engineering-oriented cultures, such as that envisioned by the proponents of the Capability Maturity Model, seek to optimize results by increasing process rigor. Proponents of an adaptive culture believe that the most critical activities—those of creativity, innovation, and problem-solving—operate best in a more flexible environment, one in which rigorous practices should be designed to balance at the edge of chaos, not to keep everything nice and tidy. My general guideline for managing larger projects is to increase the rigor applied to defining and organizing the results, and even then, *to increase the degree of rigor slightly less than just enough.*

The third solution, frameworks that encourage emergence and increase rigor, is the subject of this chapter and the next. As larger, complex projects are undertaken, there are frameworks in which rigor can be applied in ways that are not detrimental to an emergent environment. The first of these frameworks is called the Advanced Adaptive Life Cycle, which enhances the basic Speculate–Collaborate–Learn cycle by replacing the traditional emphasis of workflow with an emphasis on work*state*. The second framework is structural collaboration, which is covered in Chapter 10.

> **The key to scaling up extreme projects is to apply increasing rigor to the results, that is, to the work*state* rather than to the workflow.**

To understand business from a product perspective, we should ask two questions: "What do we produce?" and "How do we produce it?" Processes or activities are the "how." Results are the "what." Results are the components that a software development project produces. Every product, whether an oil-drilling platform or a new miracle drug, has some describable structure—a definition of the "what." Civil engineers describe the structure with blueprints, biochemists use molecular structure diagrams, software developers use use-case models. While the "what" and the "how" are inextricably linked together, focusing on one versus the other engenders different approaches to management.

Traditional software development practices, beginning with the Waterfall Life Cycle and extending to the Capability Maturity Model, are process- or workflow-oriented. The stated premise of advocates of process-driven approaches is that if the process is improved, then the product, the results, will improve also. Workflow models break the development process into increasingly detailed phases, activities, tasks, and steps (or a similar hierarchy). Upon completion, each task can be crossed off the project manager's checklist.

Adaptive development focuses not on tasks, but on the results—specifically, the primary components that deliver functionality to the end user. While documentation is defined as a support component, the components that deliver direct functionality to the end user are the primary focal point, as described in Chapter 4.

"Successful change programs begin with results. . . . Most corporate change programs mistake means for ends, process for outcome. The solution: focus on results, not activities."
—R. Schaffer and H. Thomson [1992], p. 80.

One important difference between the traditional workflow and the adaptive approaches is how each handles partially completed work products. In the workflow approach, work products are finished; in the adaptive approach, work products evolve. In a traditional life cycle, we expect the requirements document to be completed in the first phase of the project, then change slightly over the rest of the project. Change-control procedures are put in place to "control" change and hopefully keep it to a minimum.

During an Adaptive Development Life Cycle, the "create an order" component would evolve over several development cycles. While we hope the number of changes decreases from cycle to cycle, the component may not be finalized until late in the project. When the project is of small to moderate size, keeping up with constantly evolving components is not too difficult. However, as project size exceeds 5,000-to-10,000 function points, additional rigor is required to manage the project. The workflow approach to rigor is to break tasks down into finer and finer detail—often hundreds of detailed tasks. The work*state* approach is to define three or four completion "states," or status codes, for each key component. Monitoring a handful of components and a few states is easier than monitoring hundreds of tasks—and more effective, too.

A second crucial difference between the workflow and work*state* approaches is that a workflow approach assumes that we can articulate each of the tasks involved in producing the results and, furthermore, that a cause-and-effect network of the tasks can be built. A work*state* approach recognizes, particularly for the most creative aspects of development, that people may do "a little of this and a little of that," in trying to find a solution. A great developer will scratch out some requirements notes, doodle a design, code a prototype, engage in a heated discussion with a peer, and return to scratch out some additional requirements—articulating his or her sequence of creative activities is a hopeless exercise. The work*state* approach says, "Don't bother me with the detailed activities, just let me know when the work product (component) has reached a certain completion state." The rest of the chapter describes this transition to managing the work*state* rather than the workflow.

Since our management practices are so attuned to thinking about process or workflow, the next section offers some insight into breaking that workflow mindset. Then, in order to better understand the conceptual foundation of the Advanced Adaptive Life Cycle, we will explore how the information that defines a component would evolve over that single component's life cycle. The Advanced Adaptive Life Cycle is then described as a framework for managing all the components that constitute a product. Finally, the last section provides guidelines for gradually increasing the rigor of managing components as size increases. To simplify terminology, from this point on, I use the general term Adaptive Development Life Cycle to connote *both* the basic and advanced versions.

Breaking the Workflow Mindset

Software developers usually visualize phases of a software development life cycle as a workflow, because software development life cycles originated in military, aerospace, and construction projects that contained processes, tasks, and activities. This workflow mindset is so prevalent that anyone today who is not doing process improvement or process reengineering is considered undisciplined or behind the times. It's not that improving process is a wrong approach to software development, just that it has been oversold, particularly for processes with a high creative content like product development. Workflow is appropriate where definable, repeatable processes prevail. For creative work, another mindset is more productive—a results-oriented mindset.

There are keen differences between process-centered and results-centered viewpoints. Proponents of the former divide higher-level processes into lower-level processes, which in turn contain tasks—all are linked together. The processes or tasks are definable, repeatable, precise, and sometimes independent of the task's performer. Workflow proponents assume the tasks possess both sequence and dependency, as shown in Fig. 9.1.

While a workflow approach is appropriate for some things—for example, processing an automobile insurance claim—it doesn't work for product development. Many product development activities are concurrent, with partial completion and later refinement being the norm. A task stated as, "Develop a logical data model," and scheduled for three weeks of effort, doesn't describe how the task is actually to be accomplished. Developing a logical data model isn't a straightforward process. As the analyst learns more about the client's business problem, the model is refined and detail is added. Usually, the analyst collaborates with others to enhance the model. It may be distributed to other locations in a multinational company. It may be decomposed and then reconstituted in a different form. The model may appear finished, and, weeks later, someone doing a use-case model may view the problem from a different perspective and uncover new information that must be reflected in the data model. Developing a data model is on-again, off-again, iterative, discontinuous work, not a smooth flow. A workflow approach is not an effective way to manage the creative activities needed to produce a data model.

Far better, from a project management perspective, is to define the data model as a desired result, assign it to a development cycle, and let the team worry about *how*. If the team needs to consult a data-modeling reference book for development guidelines, fine—but the detail-level steps should not be part of the management plan.

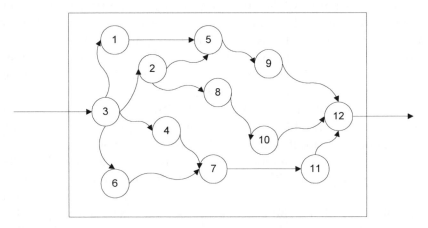

Figure 9.1: A Workflow Showing the Sequence and Dependency of Tasks.

In the context of the Adaptive Development Life Cycle, product development is viewed as a combination of creativity, concurrency, research, customer feedback, order, disorder, rework, and eventual progress. Various product components evolve at different speeds, over which development management must exercise some overall control. The Adaptive Development Life Cycle draws on concepts from the phase-and-gate approach used in developing manufactured products, in which development is synchronized and evaluated at a series of gates (at milestones and at the end of the cycle). To pass a gate, the results—a defined set of components—must meet certain performance criteria, including "state." The "state" of the component reflects how complete it is. Whereas workflow management focuses on the completion of a series of tasks, work*state* management focuses on components and their "state" of completion.

The Work*state* of a Component

Moving from a process-oriented life cycle view of software development to a component-based Adaptive Development Life Cycle view requires developers and managers to make a significant conceptual switch. To help explain this switch from task to component, this section includes a brief example showing how a single component that implements the customer credit-checking feature of an order-processing application evolves over time. We must first understand how to manage the evolution of a single component if we are going to be able to scale up sufficiently to manage a large number of components that are evolving both sequentially and concurrently.

As components evolve concurrently over time, they utilize *partially completed results* from other components. In a competitive environment in which speed is required, there isn't time to complete each component before moving on—concurrent development must occur. Furthermore, effective utilization of partial information determines whether concurrency saves time or just increases confusion and chaos.

Using Partial Information

Software developers and managers frequently face a difficult dilemma: "Do I act today or wait until tomorrow when I may have better information?" Unfortunately, in today's fast-paced business world, having full information is an unaffordable luxury. Judging how much partial information is enough to proceed with is an essential skill.

If we look at this dilemma in terms of our credit-checking example, we might imagine that the team developing the credit-checking component needs certain information about the customer-maintenance component. Relevant questions are, "How complete should the information be for us to start? How complete should it be for us to finish?" The state of a component is explicitly defined by the reliability of the information available about it. For effective concurrent development to occur, all parties have to understand how to use *partial* information. Unlike serial development in which stability comes from developers

"[Overlapping activities is] a core technique for saving development time."
—P. Smith and D. Reinertsen [1997], p. 153.

operating on information in a final, completed state, adaptive development relies on assessment of available, albeit incomplete, information.

In serial development, there are two implied information states—in-process and finished. Serial development, with its insistence on possession of complete information, has sometimes led to serious problems. For example, in the early days of Structured Analysis, there were projects known to get lost in diagramming the *current physical system* for eighteen months and more! Driven by an intense desire to know *everything* about the current business before making the transition to describe the new application, Structured Analysis proponents spent millions of dollars and innumerable man-months gathering information before even starting to build the application. Similarly, in the 1980's, many IT departments spent years trying to complete exhaustive Information Engineering plans and database designs so that application developers wouldn't have to operate using any *partial* information.

In fact, serial development provides a comfortable way to work. It helps to assure developers that no mistakes will be made from lack of complete information, and it isolates the development team from the chaos and messiness of the real world. The serial approach demands that developers severely restrict inputs and feedback because too many external changes would overwhelm the process. In development work, there are ultimately only two choices—restrict inputs or operate on partial information. Serial development seems to take the easy way out, perhaps opting to operate on full information because it appears on the surface to be less risky. The drawback is that in even mildly complex environments, waiting for full information ensures that the product won't meet the customer's needs anyway!

Concurrent development seems to be riskier than its serial cousin, but with a well-thought-out approach it is not. Concurrent development doesn't take the view that all activities can be worked on concurrently, however. For example, it would be a colossal waste of time for developers to build all the help functions into a piece of software in the early cycles, because help requirements change so frequently. Similarly, it would be unwise to begin much of the localization task of converting a software application's interface with the user into another "local" language, other than including design ideas for making localization easier, prior to having a reasonably stable user interface.

So, even in a highly concurrent development environment, there are dependencies that enable some tasks to be performed concurrently while others cannot be overlapped. The danger in overlapping is that information needed to accomplish a successor task—that is, one that depends upon the information received from one or more predecessor tasks—may not be complete, necessitating the reworking of numerous tasks. When rework causes the schedule to slip beyond the time saved in the first place by overlapping, all concurrent development advantages are lost. It is essential that all developers understand the risk of using partial information in the context of dependencies.

Figure 9.2 shows a section of a typical project plan. It indicates that Task 2 is dependent on the completion of Task 1, and specifically, that Task 2 uses a completed deliverable from Task 1. (One example from the building industry would be that the foundation must be poured prior to framing a house.) Figure 9.2 also shows a dependency between Task 3 and its predecessor Task 2. However, in this case, the overlap (Task 3 starts before Task 2 is completed) indicates that there is either an intermediate deliverable or a partial information flow from Task 2 that allows Task 3 to begin. There is always a risk associated with starting a task early—in this case, the risk might be that the partial information from Task 2 was not well understood.

"Overlapping relies heavily on the use of partial information. The effective use of partial information both requires and supports the concept of a close-knit team. Lots of face-to-face communication is essential. . . ."
—P. Smith and D. Reinertsen [1997], p. 162.

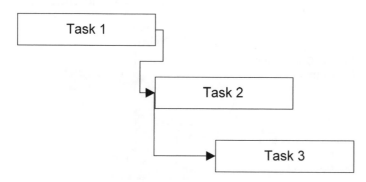

Figure 9.2: Dependent and Overlapping Tasks.

In software development, the dependency between tasks usually involves information. For example, I need a requirements specification before I can perform the design task, or I need information about the customer-maintenance component before I can code the credit-checking component. The transfer of information between the team working on one component to the team working on the next component can be formal, using documents, or informal, relying on discussions. As the information flows begin to cross organizational and geographic boundaries, both the content and context of the flows must be made explicit. Component life cycles provide both the context and content for development and for project monitoring.

Figure 9.3 illustrates several types of information transfer. Type 1, showing no concurrency, is usually associated with serial development. The group working on the successor task has virtually no information until the upstream group has completed its task. The second type provides information earlier, but the information flow is still one way. There is little chance for substantive interaction. Although the downstream group can provide some feedback, organizational barriers usually impede the meaningful interaction that leads to shortened development. Overlap of this type can cause enough reworking of tasks to negate any early start.

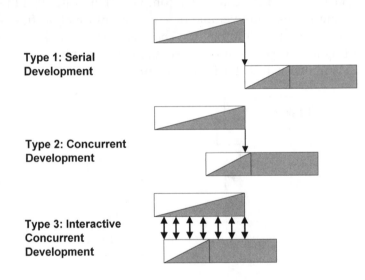

Type 1: Serial Development

Type 2: Concurrent Development

Type 3: Interactive Concurrent Development

Figure 9.3: Partial Information Flow and Concurrent Development.

While there are other variations possible, including some allowing for partial feedback, Type 3 illustrates a fully interactive, concurrent version that supports integrated development efforts. Information flow is depicted by the double-headed arrows between task cycles. Type-3 interactions are easiest in tight-knit, collocated teams. The shaded area in a task indicates the accumulation and assimilation of information. For example, in the Type-1 portion of the figure, the developers performing the first task accumulate information over time and then transfer it in a single batch. It then takes time for the developers per-forming the second task to assimilate the new information. In Type 2, the developers performing the second task can assimilate the informa-tion earlier because of the overlap. In Type 3, the developers perform-ing the successor task have absorbed the information even earlier. In addition, in a Type-3 effort, the information may be higher quality because of the interaction.

This discussion is important to managing adaptive development for three reasons. First, in order to speed delivery, the overlapping of tasks is necessary but also increases risk. In order to decrease the risk, managers must continuously monitor "completeness" of information and communicate it to the developers of successor tasks.

Second, software development is too complex an undertaking to depend on static one-way information flows. There are too many nuances that documentation cannot adequately convey. Even the serial information flow indicated in Waterfall Life Cycles works poorly in any but the simplest projects. Dynamic, two-way information flows— by means of documents, feedback, response, revision, and discussion— are the only way to reduce the risk of miscommunication in complex projects.

Third, as the "Managing Component Rigor" section later in this chapter and the "Eight Guidelines for Applying Rigor to Project Work" section in Chapter 10 will show, effective adaptive project management actually involves less administrative work than traditional task-based project management.

Component Life Cycles

Every development component, whether it is a document or a code module, evolves over time. A component's life cycle is simply its tran-sition through a series of completion states—outline, detail, reviewed, and so on. At one point in its evolution, we could declare that the com-ponent is in an "outline" state that gives potential users of the compo-nent information about its completeness. During each cycle of devel-opment, a component could be "revised" many times before it is com-plete enough to be declared in the next state. Component life cycles *explicitly* define partial information states. Understanding an overall Adaptive Development Life Cycle for the entire project is easier if the project is viewed in the context of a group of individual component life cycles.

To get a feel for component life cycles, let's revisit our InNovator product team (from Chapter 3) and examine its creation of one product component, a requirements specification document. For purposes of this example, assume that within the overall InNovator team there are two feature teams in one location and a user documentation team located five hundred miles away. Another product development team (let's call it the STAR Project) in the same company has an interest in the progress of InNovator because it will incorporate portions of the InNovator product into its own product.

Figure 9.4 shows a sample life cycle for the requirements document. During each cycle (a period of time), the document will undergo many revisions. The criteria for the component to pass the decision milestone for each cycle is whether or not it has been declared to be in a specific completion state.

Only the feature teams closely collaborating on the feature requirements need day-to-day revision information. During Cycle 1, the user documentation team members aren't interested in the requirements document at all; they are still finishing writing the previous product's documentation. In fact, while the requirements document is being developed in Cycle 1, neither the user documentation team nor the STAR team has access to it.

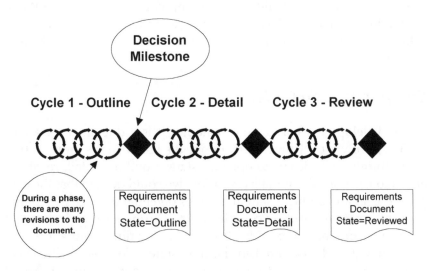

Figure 9.4: The Life Cycle of a Document Component.

At the first decision milestone, a decision must be made about the document. The question is, "Is the requirements document sufficiently complete for the InNovator team to feel comfortable releasing it in outline form?" The team leader confers with team members to assess their opinion of the document's readiness. In addition to the team's own evaluation of the document, the product marketing manager must also approve it. Once the approvals have been obtained, the InNovator team leader updates the state of the document from the initial—or null—state to outline and notifies both the user documentation team and the STAR team of the state change. When the document switches to the outline state, these two outside teams receive access rights to the document.

Rules for using shared information are developed as part of a particular team's project plan. For example, the access-rights rules governing documents in particular states are established to filter the information so that teams receive what they need at the correct time. The rules are not designed to restrict information, but rather to filter it. A detailed discussion of restricting and filtering is contained in Chapter 10.

As a next step, user documentation team members scan the outline requirements document in order to do some preliminary planning and to allocate resources, but they cannot begin serious work without more detail. STAR team members also review the outline specifications and return the document along with some suggestions that, if incorporated, would help them develop their own product.

Around the middle of Cycle 1, the InNovator team discovers the need for a major feature not included in the original document. Since the change is major, the requirements document reverts to an initial (null) state until the new-feature specifications are incorporated into the outline document and the change is approved. Returning the document to a previous state says, in essence, that it is not as complete as was first thought. The project team is therefore required to revisit the previous decision milestone. Revisiting the decision milestone ensures that all approvals are reviewed again, but even more important, it ensures that anyone who received notification of the original decision milestone is notified of the revision. Minor revisions would not require altering the component state to a previous setting, but major ones do.

No one from the user documentation team is really interested in the requirements until all the details are complete, but one member of the STAR team follows InNovator's development closely. A couple of e-mails show that the STAR representative has some really good ideas, and so the InNovator team grants him not only access rights to the document, but also annotation privileges.

At the second decision milestone, at which the document must pass from an outline to a detail state, the approvers are still the feature team members and the product marketing manager. However, the notification list of people who are to receive the detail-state document is significantly longer than was the list of outline recipients. In the final cycle, a series of technical, marketing, and management reviews are held that result in the requirements document passing into the approved state. Once the document passes into the approved state, revision rights are restricted to the team leader.

All of the rules discussed in the previous paragraphs—approvals needed at each cycle, access rights, and distribution lists—are established in the project plan. As the "Managing Component Rigor" section later in this chapter describes, the number of rules increases with project size and the team's distribution. These rules, which are used to manage the project, revolve around the components themselves, not the tasks that produce them. More specifically, the rules specify what the state changes are, who approves state changes, who receives notification when the state changes, and what access and update rights pertain to documents in certain states. The rules do not specify how to produce the components, but describe what to do with information about the components.

The biggest question on your mind at this point probably is, This is easier than task management? I think so, but it's definitely easier to do than to describe. There are two fundamental reasons this work*state* addition to the Adaptive Development Life Cycle is important. First, it is a better mechanism or framework for increasing rigor while encouraging adaptive, self-organizing, emergent behavior. Work*state* management involves telling teams what to do with their work-product information; it does not tell them in excruciating detail how to develop them. This seemingly subtle shift in emphasis is powerful, much more powerful than it appears at first.

For most product development groups today, the most difficult part of their job is not performing specific tasks, but integrating components—that is, managing intra- and inter-project dependencies. The second reason the work*state* approach is so effective is that it emphasizes information about the completion status (state) of components that in turn enhances the group's ability to manage the project dependencies.

Component Types and States

Chapter 4 introduced three component types: primary (those components that deliver function to the end user—for example, a credit-checking feature or a file-management component); technology (platforms on which the primary components depend—for example, a new database management system); and support (everything else—for example, a series of use cases). Each of these component types may use completion-state information.

In the discussion so far, three states have been described—outline, detail, and reviewed. At this point, we must define these states more precisely, add a state, and standardize the terminology. While a range of states could be envisioned, in practice, too many states are difficult to manage. Four or five states normally are adequate, although specific components may use fewer. As a standard, four states are proposed here: outline, detail, reviewed, and approved. Note that alternative state names that are used in some organizations are shown in parentheses.

1. Outline (Conceptual) State

The outline state is the initial state of completion. Part of the mission definition is considered to be an outline of the requirements. The mission statement gives enough detail to do overall project planning, but not enough for coding to begin. A code module might be considered to be in an outline state when the developers have sketched out the basic framework and some of the important algorithms but have written little of the actual code.

2. Detail (Model) State

When a component is declared in the detail state, most of the detail-level work has been done. For most primary components, this state includes a prototype or model. The component may not be complete, or fully tested, but it performs the basic functions required. A coded, partially unit-tested module might be in the detail state. A requirements document would be in the detail state when some amount of detailed explanation has been completed for each of the outline requirements.

3. Reviewed (Revised) State

All components should go through a review process. Each cycle, for example, ends with a technical review and a customer focus-group review. To pass to the reviewed state, a component must have been reviewed, and any recommended changes must have been analyzed and implemented. After a review process, if the changes were significant, the component might remain in the detail state. Elevating it to the reviewed state implies a reasonably final, tested component.

4. Approved (Available) State

The approved or available state usually involves independent verification of the component by, for example, a quality-assurance review, a cross-team review, or a management review. At this point, the component becomes available to other groups in final form. The recipient of an approved component should be able to rely on it, knowing that any further changes would be controlled and communicated.

Constructing an Advanced Adaptive Life Cycle

The previous section described the life cycle of a single component. This section describes the process of constructing the overall Advanced Adaptive Life Cycle necessary to manage all of a project's components.

Products are accumulations of individual components. While each individual component has its own development life cycle, a large group of components defines the overall development effort. Information about all the components, particularly their state, is used to monitor and manage the overall development effort. Component state provides an important tool for managing increasingly larger and more complex products.

For smaller projects, precise definition and monitoring of states is unnecessary. For example, for a thousand-function-point project, it is usually sufficient to know that a data model will evolve over the life of the project and probably will be in a reasonably advanced state by the end of the second out of four delivery cycles. With a larger project on which several concurrent feature teams are working, it may be necessary to require that the data model exist in the outline state by the end of cycle 1, in a detail state by the end of cycle 2, and in an approved state by the end of cycle 3. This additional information allows downstream feature teams to proceed with their work, but does not permit them to stray too far. Defining the state of the partial information flow enables work to be overlapped while, at the same time, it reduces the risk of excessive reworking.

Discussed first in Chapter 4, the frequent, sometimes daily, build process pioneered by independent software vendors (ISVs) provides a good example of how to synchronize a product's components. Each build forces developers to test the readiness *state* of their code by running it together with all other components. But while frequent builds are critical to success, they are insufficient in cases in which complexity and dependencies require synchronization on more than completed code components. Teams also must be able to synchronize on partially completed deliverables. Each decision milestone is a synchronization point for components under development.

The concepts behind the Advanced Adaptive Life Cycle, specifically those related to monitoring and managing projects by means of component states, come from the phase-and-gate development approach used by a growing number of companies to oversee product development. The phase-and-gate approach stresses managing the "what," not the "how." It focuses on identifying and planning components, determining the dependencies and interrelationships between components, monitoring the evolution of each component through defined completion states, and evaluating and monitoring progress by means of both fully and partially completed components at the end of each cycle.

Phase-and-gate development is a technique for scaling up an iterative approach to address larger, more complex, high-speed, high-change products. Kodak established the phase-and-gate approach in the early 1990's to manage development of such diverse product categories as chemicals, microfilm equipment, copiers, and cameras. Their stated goal was to shorten product cycle times, improve development efficiency, and improve inter-group collaboration.

Phase-and-gate development provides a framework of task-free phases separated by decision gates. The key distinction of a phase-and-gate life cycle is the evolution of components, through a series of completion states, to final approval. In terms of the Adaptive Development Life Cycle, phases are equivalent to cycles and gates to milestones.

"Task-oriented workflow should enter a product development model only after a task-free Phase and Gate backbone is firmly in place. The modern phase and gate techniques in concurrent development management resolve the conflict between accountability and visibility on one side and spontaneity on the other." —W. Collier, D.H. Brown & Associates, private correspondence [1996].

Cycles (Phases)

In previous chapters, we have used the term "cycle" but we need to further define it in the context of this discussion. Cycles are time periods during which some part of the development effort occurs. Although similar in general appearance to more traditional project phases, cycles are periods during which concurrent activities occur, either in separate organizations or in integrated cross-functional teams. As was stated in Chapter 4, software components are iteratively developed over several cycles.

During a cycle, individuals and small teams are bound not by specific tasks but by a common understanding of the desired results. Each cycle has a distinct set of aimed-for components and although there may be an initial conceptual development effort, by the second cycle, a model or prototype should emerge.

The phase-and-gate approach, like Adaptive Software Development, promotes the delivery of interim tangible prototypes rather than documents. In the electronics field, the model may be a breadboard version of the product; for large products like automobiles, it may be a scale model; for shoes, it may be a hand-tooled mock-up. Whatever the field, the important point is that the product development system produce a series of tangible, visible, testable renditions of the final product—not merely descriptions on paper.

Figure 9.5 shows the flow of components during cycles of product development, like fish swimming past the aquarium observer. Some components start and finish during a cycle, while others overlap two or more cycles. Although not shown in the figure, some may start early and lie dormant until much later.

Figure 9.6 provides another view of the components' evolution, indicating dependencies. Four states are shown for the primary component, but only two states are needed for the technology component. It is not necessary to recognize every state for each component. In Fig. 9.6, the technology component might be a LAN network, the primary component a software feature, and the support component an object model. The diagram shows in graphic form the dependencies of components on partial and final states of other components.

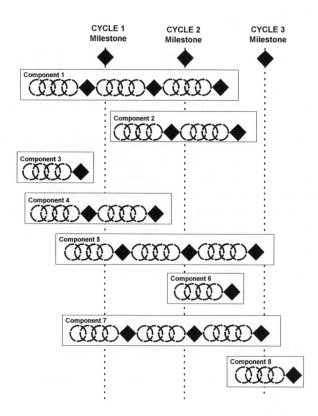

Figure 9.5: Components in an Adaptive Development Life Cycle.

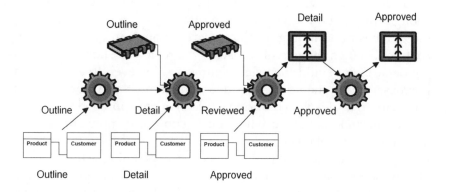

Figure 9.6: Component Dependencies.

Milestones (Gates)

Periodically during the development process, milestones are established. A milestone or gate is a decision point, much like a traditional project management milestone. For the most part, a gate and a project milestone are synonymous. However, in traditional task-oriented project management, milestones are more typically related to task completion rather than to components and component states. Milestones are synchronization points, at which the results from ongoing efforts are brought together, reviewed, or integrated into a testable product. The decision as to whether a project passes a milestone depends on the list of planned components for that milestone and their specified completion state. For example, the first gate for a PC component product such as an Ethernet card might include

- a breadboard model of the primary circuits

- a detailed, overall-design document

- an outline-level integration test plan

- a conceptual or primary-level drawing of the packaging

- an outline-level blueprint (rough draft) of the manufacturing floor

- an outline-level (preliminary) parts list for purchasing

In the above list of sample components for the PC peripheral card, states are included in the descriptions. For example, the parts list is in an outline state, which provides the purchasing department with enough information to begin looking for parts suppliers and to begin the purchasing planning, but not nearly enough information to sign contracts. When the parts list reaches the reviewed state, the purchasing department could then sign contracts. However, to actually issue purchase orders, the purchasing department must also have a production plan (to be developed during Cycle 2) in the approved state.

A milestone is passed when a collection of components is determined to be in a planned state (that is, all components in the collection were planned during the initial project-cycle planning activity). A single component, such as a feature set, must pass certain prescribed test criteria before it changes state. Similarly, for the entire product, a series of components must be in specified states, as shown in Fig. 9.5. Milestones provide opportunities to monitor progress from a management perspective. Looking again at Fig. 9.5, one can see that components at different completion levels can define a milestone. For example, Component 4 (a particular feature set) would have to be in an approved state as a condition for passing the second milestone.

Milestones allow us to take the overall pulse of the development process and to assess its health. However, they are not intended to stop development while the review is conducted. Development of many components can, and should, progress during the milestone review process. Milestones are intended to monitor progress, synchronize components, and obtain approvals to proceed, not to become a roadblock. They may also be used to open the development results to a wider audience.

Milestones are critical to success. Without well-defined points of synchronization, the product development process degenerates into a chaotic mess. There is a big difference between operating on partial information in a known state and operating on information of unknown completion status. Knowing that a set of components has passed a cycle milestone and that a number of people have reviewed its progress, provides development teams with a clearer sense of what should and should not be done with partially completed components.

Managing Component Rigor

Some organizations are beginning to substitute an adaptive, results-oriented life cycle in place of a workflow approach. They are managing *components* rather than *process*. However, there is a third leg of this conceptual triangle—one that is often missing. That missing piece is managing results. Having relieved themselves of the bureaucracy of process management, organizations avoid *all* rigor and thereby doom themselves to the twin demons of complexity and size. Managing the *results* in larger projects necessitates increasing rigor and discipline. It also requires increasing automation. Unfortunately, most of the industry's automation tools have been developed to support workflow-type life cycles. For example, while many configuration control and collaboration tools provide version control, few have a direct way of providing state information.

As the size and complexity of products increase, three basic strategies can be employed:

1. Increase component rigor.
2. Increase emphasis on defining and monitoring dependencies.
3. Refine state transitions.

A fourth strategy is to support the other three strategies with increased levels of automation. Chapter 10 provides ideas on the automation of work*state* management through the use of what I call a *collaboration service layer*—a set of tools for structural collaboration. The following sections detail the three basic strategies.

Increase Component Rigor

Project managers often equate good management with micro-management. Their project work-breakdown structures (the list of all project tasks) burden the project team with daily activities and hourly "to do" lists. These managers forget the distinction between doing work and managing the project. Detailed task lists convey a not-so-subtle message to the project team, "I don't trust you to do the job correctly."

A similar obsession with detail can adversely impact adaptive management. Just because a product can be subdivided into one hundred components doesn't mean it should be, at least from a project management perspective. While configuration-control applications are usually directed at managing all components, project management needs to focus on essential components. For a five-hundred-function-point project developed over three months, it should be sufficient to allocate the primary components (possibly defined by a two-page requirements outline) to three cycles and to control the project by means of regularly scheduled focus groups and technical reviews. In this simple case, there is no need to break the feature components into more detailed subcomponents or to create monitoring states.

As the size increases, say, to a thousand function points, and there are other groups who must use the requirements document, there might be a need to plan for three states for the document—outline, detail, and approved—in order to reduce the risk of sharing partial information and to adequately monitor the project.

The rigor of managing components can be increased in four ways: by dividing components into smaller subcomponents in order to manage smaller chunks of work, by imposing intermediate completion states on components, by increasing the formality of a component's documentation (from scratches on a napkin to a word-processing document, for example), and by increasing the number of components monitored (all components must be developed, but not all components need to be used to manage the project).

Increase Emphasis on Dependencies

With smaller projects, dependency relationships are informal and remain implicitly understood. When teams are small and communication informal, project members just seem to *know* when to start work on developing the next component. Occasionally, something slips, but the correction time is usually short. Bigger problems arise from forgetting a component that was not explicitly identified up front, not from working on components out of sequence.

As project size increases, and as the number and geographic dispersion of feature teams increase, defining and monitoring component dependencies becomes a critical management issue. As components are developed at greater physical and organizational distances, informal communication needs to be formalized. (To understand the impact of physical and organizational distance on teams, think how, when two teams report to the same manager, their organizational distance is narrow. When these same two teams' manager's manager's manager reports to the same person, their organizational distance is wide.)

In traditional project management, a dependency defines a cause-and-effect relationship between two tasks. One task must be completed before the dependent task can be started. While at a code-module level there may be a physical dependency (the credit-check module depends on the customer database), many of the dependencies in a software project are informational—for example, a systems designer needs requirements information prior to design, but as we saw earlier in this chapter, accelerated projects must utilize partial information.

Project managers should carefully identify dependencies, establish adequate monitoring processes, and improve their problem-anticipation skills. An unidentified dependency within a single, collocated team is usually identified informally—such as during a lunch-time conversation when one developer suddenly realizes another team member possesses some information he or she needs to know. However, with dispersed teams, when one team is dependent on information from another team, and the dependency is neither identified nor communicated properly, the chance of a problem is heightened because there are fewer opportunities for informal resolution.

In larger projects, intergroup dependencies are rarely given sufficient emphasis by management. One reason may be management's failure to understand how component relationships impact one another. By focusing on the critical components that cross organizational boundaries and by monitoring critical states of completion, managers can become more effective with less overall effort. Management emphasis is thereby moved from monitoring of tasks to monitoring of components and their interdependencies.

For independent software vendors, the problems caused by these dependencies extend beyond producing code. ISVs must schedule efficient use of manufacturing facilities, print manuals, localize the product (that is, translate documentation and applications into the native language of the targeted customers), arrange sales training, plan product launches, and more. Poor dependency management acts like a tsunami breaking over groups dependent on the availability of good information. What seems to developers like a minor swell in the ocean becomes a tidal wave as it approaches the shore of product shipment. Managing tasks well may be important, but managing dependencies well is critical.

Refine State Transitions

For a simple project, such as the three-month five-hundred-function-point project mentioned earlier, the manager can adequately monitor the project's components using two states—Not Done and Done. At the other end of the project-size spectrum, each of the hundreds of components in a product might require four or five states for adequate project control. Refining state transitions is part of the project planning process in which the project team identifies critical components and the required number of states for each. For example, components used entirely within a collocated team will need fewer identified states; components with many interlocking dependencies will need more states. The need for intermediate states is tied to dependencies between components and to the need to make partial information states explicit.

Managing Workflow in an Adaptive Environment

From the discussion in this chapter, it might appear that any use of workflow is discouraged. This is not the case at all. Workflow—that is, the processing of information through a series of well-defined activities—is extremely important when properly used within an adaptive life cycle. For example, the daily build process needs to be extremely detailed, precise, and rigorous. Without this rigor, further testing and development can become a configuration nightmare as application size increases. The daily build process is an example of a necessary workflow process.

The workflow approach drives many traditional software engineering activities, and is appropriate in adaptive development when the workflows are *additions* to the adaptive framework—they are not the *primary* life cycle framework. That is, workflow is a piece of the solution but not the structure. Some activities may have a very well-defined, rigorous process workflow while others may have more flexible workflows; many may have no workflows at all. Some workflows will be discretionary to each autonomous group; others will be required.

In my view, the Software Engineering Institute has useful practices to develop and manage projects (configuration control, risk management, and so on), but the Capability Maturity Model is just not the right organizing model or framework for adaptive projects. The CMM is a workflow model, based on the belief that complex products can be effectively attacked by breaking a large process into ever-more-refined subprocesses. An adaptive model utilizes many of the same components as are advocated by the SEI, but from within a very different culture and by taking a radically different approach to overall life cycle structure.

Summary

➢ The secret to managing complex adaptive projects is to apply increasing rigor to the results, not to the process.

➢ Using a results-oriented framework for collaboration enables managers to better balance the issues of rigor and flexibility. It provides an optimal blend that allows for creation of large, complex products without stifling the innovation and creativity needed to operate at the edge.

➢ The transition to managing results involves replacing the workflow mindset, focusing on specific result components, and then applying new ways of managing those results.

➤ The workflow mindset is process-oriented first, and results-oriented second. Work*state* management reverses the order of importance, making management of results the focus.

➤ Work*state* management is well aligned with how complex adaptive systems operate.

➤ One of the key factors in scaling up concurrency and iteration is the use of partial information. Using information *state* characterizes a component-oriented information life cycle.

➤ There is a tremendous difference between using serial/batch and interactive/complex information flows. The latter supports concurrency and collaboration, but it also carries higher risk.

➤ Phase-and-gate life cycles, the conceptual base for adaptive life cycles, are characterized by an emphasis on results. They help resolve the conflicts between innovation and accountability.

➤ Milestones are important decision points. They explicitly address transitions between partial information states and establish guidelines for allowable actions on information in each state.

➤ Managing component rigor emphasizes increasing the rigor applied to component *context* information as size and complexity increase.

CHAPTER 10
Structural Collaboration

At Sun Microsystems, the slogan "The Network Is the Computer" implies a high degree of communication among collaborating staff, but a network alone does not create collaboration. The World Wide Web ferments with innovation and anarchy. New products, new markets, and new demons have emerged from this polyglot of interconnected people and companies. And while software product development benefits from an emergent environment such as that on the Web, it needs to nestle closer to the orderly side of the edge of chaos than to the anarchy side.

Two themes of this book have been that emergent order is necessary to solve complex problems and that collaboration produces emergence in organizations. Furthermore, the most effective environment for generating emergent results is one that balances on the edge of chaos. "Connect everyone to everyone and let creativity flourish" seems to be the mantra of many who see the Web as building a new social and organizational order (Kelly98). But connecting everyone to everyone seems as likely to create chaos as order. If biologist Stuart Kauffman's simulations (examined later in this chapter) and Ralph Stacey's work on networks of people (introduced in Chapter 5) are indicators, then structuring an effective collaborative environment is more difficult than giving everyone a browser and an e-mail account—

"The two forces that we have always placed in opposition to one another—freedom and order—turn out to be partners in generating viable, well-ordered, autonomous systems. If we allow autonomy at the local level, letting individuals or units be directed in their decisions by guideposts for organizational self-reference, we can achieve coherence and continuity. Self-organization succeeds when the system supports the independent activity of its members by giving them, quite literally, a strong frame of reference."
—M. Wheatley [1992], p. 95.

it goes far beyond the typical communications network that merely connects people.

As the size of a product-development organization increases beyond a single feature team, the difference between merely sharing information and sharing creation becomes more pronounced. Knowing where to go for information, knowing who knows what, and knowing how to influence events become increasingly difficult. The maximum size at which informal mechanisms support an adequate level of collaboration and sharing is about two hundred to three hundred people, according to the research of Thomas Davenport and Laurence Prusak (Davenport98). Their findings, combined with Lipnack and Stamps's research (Lipnack93) that indicates collaboration deteriorates at distances exceeding fifty feet, makes it obvious that strengthening structural considerations—such as practices and processes, organizational structure, linkages to knowledge management, and technology infrastructure—is critical for most organizations desiring to improve collaboration. As stated previously,

> **The task of structural collaboration is to deploy methods and tools that support self-organizing principles across virtual teams.**

In the turbulent markets of today's e-business economy, a project team's ability to adapt quickly to external changes is fundamental to its success. Teams that have both the authority and the capability to process disparate information and make effective decisions are strongly self-organizing. They create a local node—a self-sufficient unit or work group—that is adaptable. A coherent network (that is, one driven by a common mission) of many adaptable nodes creates a larger but still adaptable organization.

Practices and processes provide the framework in which collaborative sharing occurs. Some frameworks, waterfall development for example, are less conducive to sharing than others such as RAD or prototyping. Rigid processes are more likely to restrict the free flow of collaborative interaction. As we saw in Chapter 8, rigid, hierarchical management organizations are less conducive to collaboration than are horizontal, team-based structures.

The work*state* practices described in Chapter 9 are intrinsic to structural collaboration. The structural component of collaboration is built on interpersonal and cultural foundations. Structural, interpersonal, and cultural ideas are necessary for tackling larger, complex projects and for accomplishing the goal of *providing a path* for organizations that need to use an adaptive approach on larger undertakings. Work*state* practices assist in managing the dependencies among projects that may need to share components, people, or knowledge. Managing these interactions among multiple feature teams across multiple products is difficult and, as my colleague Lynne Nix laments, "even companies that are good at intra-project management often fail to understand the extra resources needed to manage the dependencies between projects in complex environments."

Meaningful product-development interchanges contain information, and sometimes even knowledge. While there is an overlap between the subjects of collaboration and knowledge management, they require different types of expertise. However, at some level, they need to be integrated or linked together—part of the work of structural collaboration. Knowledge management, although beyond the scope of this book, provides methods for sharing tacit knowledge (inside one's head) and explicit knowledge (documented information). Knowledge management techniques for sharing expertise and best practices should be incorporated into structural collaboration.

Finally, the technology infrastructure impacts an organization's ability to collaborate across time and space (both geographic and organizational). While implementation of a product like Netscape's Collabra does not *guarantee* a good, or even a minimal, level of success in the collaborative effort, it does provide an enabling structure without which virtual teams have little chance at effective collaboration. The technology infrastructure includes both platforms (for example, Lotus Domino) and specific tools (for example, Microsoft NetMeeting).

In this chapter, there are four very important points made about structural collaboration. The first deals with differentiating between context and content and emphasizes the importance of contextual information. Understanding the difference between context and content is important to building an effective collaboration structure.

The second point of this chapter is to identify services and tools that are helpful in building collaborative structures. A section on larger project teams and virtual teams provides a backdrop for the discussion of these tools and services.

The third point of the chapter is to revisit the concepts of collaboration and emergence. The topics provide some design guides for building an emergent collaborative structure. The tools available for creating networked organizations provide an often-bewildering variety of possibilities. The real challenge is how to *tune* the network into an *effective* collaborative structure. This is one area in which research into complexity theory will pay significant dividends. By simulating complex phenomena and striving to understand just what "the edge of chaos" means in organizations, we should be better equipped to design and tune collaborative networks.

The fourth point of the chapter is to provide a set of guidelines for applying rigor to adaptive projects. Chapters 9 and 10 together present an array of concepts and practices for scaling Adaptive Software Development up to larger projects, but the final section of Chapter 10 provides eight guidelines to help developers determine just how much rigor to apply to a particular project.

The Critical Distinction Between Content and Context

In 1966, Eliza was born. Developed at the Massachusetts Institute of Technology by Joseph Weizenbaum, Eliza was the first computer program capable of carrying on a conversation (Leonard97). Eliza's rise to fame was meteoric, particularly in the context of her persona as a Rogerian psychotherapist. Designed to mimic the Rogerian approach, which helps patients clarify their thinking by turning all conversation back to focus on the patient, Eliza appeared to be a thoughtful human listener. For example, a patient's comment, "I feel distressed," would be countered by Eliza's response, "Why do you think you feel distressed?" By limiting the context of the conversation to the Rogerian model, Eliza was able to attract a significant number of "patients."

Researchers into natural language, and particularly those connected with the Artificial Intelligence community, continue to study ways to understand language's dependence on the context. For example, when a person is thrust into a new situation using a new technology, a new business function, or into a new market, much of the learning involves establishing new contextual information. As Eliza proved, within a well-defined context, a little data can have a major impact. Eliza didn't "know" much, but she used what she knew very effectively. Unfortunately, a lot of data with little context have the opposite effect: The data are wasted.

If we are to generate emergent outcomes from strong connections between diverse groups, the balancing of content-laden information with context-rich information is a key design consideration. One of the reasons for the success of RAD projects is that their team members are dedicated full-time to the project and are geographically located together. Despite having minimal formal documentation, RAD projects succeed because of highly interactive, team-member communication for which the context is well understood. Working papers, whiteboard scribbling, documents and drawings to create and mark up, bull sessions, hallway exchanges, and a large measure of trust, all contribute to the context necessary for successful collaboration. Team members know who produced what information, how complete it is, what should or shouldn't be done with it, who needs to look at it before it gets distributed more widely, and more. These contextual issues are nearly transparent when team members work in close proximity. As organizational and geographic distances increase, context is lost, often without team members even suspecting or understanding the magnitude of the loss. Context must be restored or the effort of shared creation becomes too difficult.

Imagine a situation in which you receive an e-mail message with no author, no subject, and no date, containing a message peppered with unfamiliar technical terminology. Or, consider needing to review a hundred-page document without having an idea of what the latest revisions were. Contextual issues such as these hinder collaboration and therefore hinder progress of any kind!

Imagine further that you arrive at your desk one morning and find a specification from another product group within your company. Your product team needs to use this particular application-program interface (API) specification. After looking it over, you jot down several questions:

- What is the status of the API specification? Is it a draft? Is it final and approved?

- Who produced the document? Are there others using the API? What confidence level do we have in the developers?

- Are the developers open to suggested changes?

- How up-to-date is the specification? Will updates be sent on a regular basis?

- Are there any reference documents that would help us understand the specification?

"Many communications theorists separate the content of a message from its context. They point to the metamessage— the relationships, status, and interpretive cues that ride along with the literal symbols themselves."
—J. Lipnack and J. Stamps [1997], p. 86.

- Were other solutions, methods, or options considered? If rejected, why?

- Is this the whole story? Are there some other things, not in the document, important to successful use?

- How closely will the final code match the specification?

These are questions about *context*. The document itself could be perfectly clear, but its context fuzzy. Context is a combination of *informational* and *interpersonal* elements, as shown in Fig. 10.1. The *informational* elements of context include those missing from the e-mail of the last paragraph—producer, timing, status, background, related information, definition of terms, version, and permissible actions. Or, to use more general terms, identification, revision, state (covered in Chapter 9), and relationships. The *interpersonal* elements of context include trust, respect, participation, commitment, and shared responsibility (these elements were addressed in Chapter 5). The question for adaptive management is how to foster these qualities across space, time, and organizations.

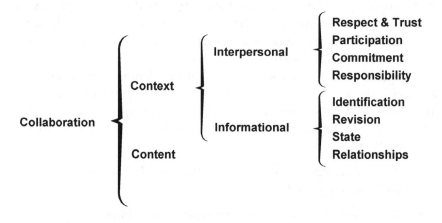

Figure 10.1: The Elements of Context.

Software development projects create components—code, test plans, requirements documents, data models, project plans, vision statements, sketches on napkins, e-mail messages, notes, pizza orders. For successful collaboration to occur, both *content* and *contextual* information must be elaborated. As size and complexity increase, contextual information must be made explicit.

Identification information includes items such as component name, project identifier, author, and feature team. Revision information includes all version, revision, and release data, as well as details of the change history and who made the changes. For word documents, the version, redlining, and annotation features already contained in word-processing programs are sufficient for many projects. An important issue with revision information is the consistency (or inconsistency) of information and its representation among various tools—CASE tools, configuration-management tools, word-processing tools, and graphics tools, for example.

State, as explained in Chapter 9, defines the status or usability of a particular component. For a specific component, state data might include current state, state history with dates, state template (that is, which states are in the component's life cycle), and estimates of the date of the next state transition.

Relationships define how individual components are related to each other. Simply placing documents related to a feature set's requirements into a common directory implicitly creates relationship information. Five types of relationships are important to collaboration: structural, reference, classification, dependency, and who-and-what.

1. *Structural* relationships define how information elements are organized; for example, indicating that two code modules are part of a particular application or that a data model is part of a specific requirements document. These structural relationships are often maintained by some form of configuration- or version-control system. The tools for managing this type of context information are widely used.

2. *Reference* relationships point from one piece of information to another. An example of a simple reference relationship could be found in a design document with imbedded Web links. In the body of the document, these might be links to other feature teams' documents or to a user-interface design document.

3. *Classification* relationships are used to create topic hierarchies; each component needs the ability to be assigned to multiple classifications. Different hierarchies help team members by offering alternative organizing and searching criteria.

4. *Dependency* relationships are important for project management. Like task dependencies in project-management tools,

component dependencies create a PERT-chart-like network that impacts the order in which things can be developed. Concurrent development does not mean that everything can be developed at the same time.

5. *Who-and-what* relationships are those between components and the people who create and manipulate those components. These relationships answer the basic questions of *who* has access to *what* components. This chapter discusses who-and-what relationships in detail.

The distinction between context and content is critical to understanding structural collaboration. Although many configuration-management tools maintain relationships between components, explicit differentiation between context and content is often missing. Unfortunately, context is not as well-defined a concept as content, as can be seen in the similarities between the definitions of the five types of relationships above. Structural collaboration defines the framework (people, technology, practices, organizational design, and more) that assists large, distributed groups to function more effectively. Structuring content is relatively straightforward; structuring context is not.

Collaboration Services and Tools

Developing a collaboration framework involves understanding the characteristics of virtual, distributed teams, providing good tools that are supportive of collaborative processes, and creating new roles that facilitate collaborative interactions. The next sections address each of these efforts.

Large Projects and Virtual Teams

If virtual, distributed teams are to have any chance of delivering products in extreme environments, the issues related to building a *network of teams*—that is, groups of interconnected feature teams working toward a common goal—must be resolved. The 1990's saw an explosion of interest in teams and teamwork, but most of the focus then and since has been on small, collocated teams. Finding solutions to the issues surrounding virtual teams is next on the horizon.

Feature teams are the basic building blocks of larger project teams. They are *core teams,* whose members (generally five to ten in number) perform the bulk of the work. The core team may be supplemented by part-time staff, called extended-team members, and by subject-matter experts for special needs.

A product team can consist of one or more feature teams. All the participants in the product team have the same overall mission, with each feature team having a specific part of the mission. Although interaction across feature team boundaries is harder than internal inter-action, the common mission provides a mechanism for cooperation. A product team is therefore a network of teams, or what Lipnack and Stamps call a TeamNet (Lipnack93).

> *Unlike conventional teams, a virtual team works across space, time, and organizational boundaries with links strengthened by webs of communication technologies.*
> —J. Lipnack and J. Stamps [1997], p. 7.

Within large organizations, there can be multiple product organiza-tions, each with multiple feature teams. Often, although the product teams have different missions, they also have interdependencies. For example, one team builds the communications module that a second team will incorporate into a product. Dependencies cross company as well as team boundaries. Teams, therefore, have multiple kinds of cross-boundary interactions. Managing them is not an easy task.

Building dedicated and collocated collaborative feature teams is difficult enough. Building collaborative, virtual team networks is even more difficult, but the difficulty is not an issue of technology. It is a matter of social, cultural, and team skills. According to Lipnack and Stamps, "It's 90 percent people and 10 percent technology. Social fac-tors above all derail the development of many virtual teams" (Lip-nack97, p. 168). Software developers, in fact many technical groups, are not noted for their team skills, but if they are to succeed in extreme environments, team skills are as vitally important as technical ones.

"On the whole, a vir-tual team must be smarter than a con-ventional collocated team—just to survive."
—J. Lipnack and J. Stamps [1997], p. 189.

Team members spend comparatively little time during the total project thinking about the network of people with whom they need to interact over the course of product development. They think that as long as they have an e-mail alias list that has been set up by some cen-tralized network administrator and a quickly generated list of people to be copied for status reports, they are ready to tackle the "real" work.

People invariably complain about the lack of communication, the flood of unrelated e-mail, and the overabundance of meetings, but they somehow do not connect these problems with their own lack of emphasis on carefully planned or properly identified communication and collaboration channels. Their reaction to poorly run, unproductive meetings, for example, is to schedule *no* meetings, instead of trying to improve the quality of the meetings.

Reducing unrelated information clutter should be an ongoing project management activity. There is, of course, an easy way to handle the problem—just send everything to everyone and let the recipient sort it out! Easy, but not effective, and so we need guidelines for sharing information. Reduced to simplest terms, these guidelines establish practices to distribute, filter, and restrict data.

Team self-organization and emergence depend on rich information flow; therefore, the first step is to encourage team members to *distribute* relevant information widely. Sharing, not controlling, is the highest-priority objective.

At some point, even the sharing of relevant data causes overload and clutter. While we may want access to another team's documents, we may not want to be notified about each and every revision. *Filtering* should be designed to increase the effectiveness of information flow by reducing clutter. It is not intended to keep secrets or restrict information.

There is, of course, some information that must be *restricted:* Sensitive competitive and financial information, long-range plans, or details about advanced technology all may need some degree of restriction. However, restriction should not be used as a tool to combat lack of trust. If some group uses information inappropriately, the solution is to make that group accountable for that use of the data, not to restrict the data. Keeping development information from, say, the sales department reduces a source of important feedback and damages the relationships between groups. Restrictions used because one group doesn't trust another leads to reciprocal actions, discouraging the very flows that enable collaboration.

Organizing and maintaining an optimal network of people—people who work well together—is critical to project management success, but what characteristics should one consider when building this strong team network? Four kinds of communications capabilities are critical:

- nodes and links

- organic growth

- push and pull

- who and what

Nodes and Links

Nodes and links define the structure of the communications network at a logical rather than physical or technological level. Nodes are the end points in a communications network and, for our purposes, can be defined as a subset of one or more individual team members. As the communications network is set up, a node could be a feature team, a subgroup within a feature team, or an individual feature team member. Links are the connections that allow information to be transferred between nodes.

For the hypothetical InNovator project described initially in Chapter 3, the state-management feature team consists of three developers, two testers, and a team leader (call him Walter). These six individuals will constitute a node in the InNovator project's team network when that network is established. The state-management feature team is tasked with developing the portion of the application that monitors component states. A second feature team, the version-control team, works on maintaining versions of documentation. However, since the work of the version-control feature team is so closely tied to state management, one of the version-control team members (call her Sherry) has actually been assigned to the state-management communications node.

An individual can be a member of multiple nodes. For example, Walter is a team leader for the state management team, and he is also eligible to be a member of the team leaders' node, in which team leaders (from both inside and outside the InNovator project team) share their knowledge and experience. Sherry is a member of both state-management and version-control nodes, and is also a member of the product team's part-time architecture-support node. Sherry acts as both the source of architecture information for the state management team and as a collaborator in the overall development of architectural guidelines for the whole InNovator product team.

Nodes in a collaboration network are varied. Each node is populated by individuals who work together in order to accomplish some goal, whether that goal is developing a particular product feature or sharing project management information.

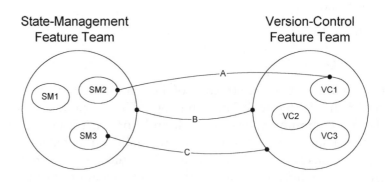

Figure 10.2: Nodes and Links.

Links are the permitted connections between nodes. In Fig. 10.2, e-mail connections are diagrammed, showing that linkage A might be a simple e-mail link between two members of the feature teams. Linkage B might indicate that e-mail between the two teams would be delivered to all members of both teams. Linkage C might indicate that while SM3 (call him George) is not a member of the version team, he receives *all* the version team's internal mail (maybe he is the designated technical liaison with Walter and Sherry or the version team).

Links can be both horizontal (for example, links that cross from one development team to other development teams) and vertical (such as hierarchical links to various administrative and product management nodes). However, it is not only the links but also the dimensions of the links that are important. Dimensions such as Ralph Stacey's control parameters introduced in Chapter 5—rate of information flow, degree of diversity, richness of connectivity, and so on—provide a basis on which to fine-tune networks for effective self-organization.

In network communications, a measurement of rate of information flow is bandwidth—higher bandwidth increases the total amount of information that can be exchanged. The concept applies to networking teams in Adaptive Software Development projects as well as to communications. Richness of connectivity measures the number of connections between nodes—for example, ten nodes, each of which is connected to only one other node, would have low richness whereas having all the nodes connected to each and every other node would indicate high richness. Degree of diversity measures the type of information that can be shared—from the simplest e-mail message to complex three-dimensional product representations and videos. A good example of both richness and diversity occurred when one of my clients upgraded its telecommunications links to Europe and Asia in order to provide both higher data volume and richer information exchange (graphics, drawings, sound, and other data) for collaborative product development.

While this discussion of nodes and links may seem somewhat off the topic, it illustrates how groups actually work, whether the links are formal or not. When groups are small and all members are collocated, the links can be implicit and informal. As groups become distributed, the links must be made explicit. The establishment of explicit links requires thinking both about whom to connect (nodes) and how to connect them (links).

Organic Growth

Self-organization—the capacity for independent agents to work together to produce emergent results—is successful when there is a well-articulated mission and the agents are given enough freedom to innovate. Self-organization fails when order, structure, and control are imposed by Command–Control management. Similarly, an imposed collaboration structure can thwart self-organization of larger groups. The question is, "How do we *provide* structure without *imposing* structure?" The answer is to follow an organic growth approach, one for which the feature teams themselves manage their own collaborative structure.

Many collaborative tools are designed around a security model— that is, the tools control how people are connected and what information they can manipulate—rather than around a service model in which the participants themselves establish the connections and information-access rules. Using a security-driven approach, integrating a new member into a network or a team, creating alias lists, defining links, and assigning access rights are usually centrally controlled by a network administrator. While restricting access is necessary at times, it is not as important as enabling the participant groups to manage network growth according to their own needs.

To create a self-organizing environment, the network must support organic, bottom-up growth. In an active collaborative environment within an organization, there are far too many nodes, links, and changes to establish timely central control. In the organic model, minimal controls are established at a high level (for example, by determining who can establish a project team node), and then control migrates outward (for example, a project team leader could designate who has authorization to establish the feature team nodes). An overall security administrator might be responsible for adding individuals to the network, but project and feature team managers would control which staff members were assigned to their teams.

An organic network grows and changes from within, under local control. The Internet is an organic network: There are a few centralized functions such as domain-name assignment, but there is no central control. An organic-growth model, or framework, supplements the basic security features of a network. Using this model, each team establishes itself, manages the access rights to its own data, and negotiates links to other nodes.

We can use the InNovator project for illustrative purposes again. Imagine that the InNovator product manager is granted authority from the vice president of Development to establish a high-level product node on the network. Except for privileged information, the product manager grants authority to feature-team leaders authorizing them to establish their own nodes. The feature-team leaders then determine what information should remain private within each feature team and what information should be public. They do this not because they are trying to be secretive, but because they are trying to reduce information clutter and eliminate extraneous information. In this process, they might assess which members of the state-management feature team need to see which pieces of communication from, say, the version-control feature team.

InNovator's feature-team leaders create team work space, assign members to the team, determine who has access rights to team data, and determine who outside the team can have access rights. Imagine a scenario in which George, a member of the state-management feature team, works on a component similar to a component being built by the version-control feature team. He asks the version-control team members to include him in their discussion forum on the component. The request is discussed, agreed to, and easily implemented by the team leader or group facilitator.

Team composition also changes over time. Within a larger team, individual feature teams may find themselves re-forming periodically. If teams constantly adjust the information-access relationships they have with other feature teams and with other product teams, and if the membership of teams themselves constantly changes, then maintenance of the network must reside at the team level.

Push and Pull

One of the new project roles is that of a collaboration facilitator (described in greater detail later in this chapter). A collaboration facilitator's job is to "think" about communications rather than let it happen by default. Part of the thinking task is determining what information should be sent automatically to team members (a push) and what information should be made easily accessible at a team member's request (a pull).

Again, an example illustrates the point. For the sake of simplicity, let's say that the InNovator project team's communications tools consist only of e-mail and a Web page. The InNovator product team has its own Website and each feature team has a linked subsidiary site. Each site has three levels of access—private, public, and restricted. Private data are accessible only to a node's members. Public data are available to anyone in the company with a Web browser. The public area contains summaries, reference information, the feature team's mission statements, status reports, and all general items of interest to a wide audience. Restrictive data are available only to the InNovator team.

The InNovator project uses both *push* information (an e-mail with an attached project status report, for example, which can be sent to a selected group each month) and *pull* information (information that can be accessed by someone on another product team who is not a regular recipient of the project's status reports but who can log on to the Website to see how the project is going). Part of the collaborative planning process is to determine which information should be pushed and which should be made available to be pulled.

Who and What

Another consideration in building a team network is to determine who should see what. In other words, what are the access filters or restrictions on access to specific data by specific individuals, roles, feature teams, or groups?

Although totally impractical, the simplest way to link components and people would be to allow everyone all actions on all components. Conversely, every component could be associated with every individual for every possible action (create, add, modify, delete, publish, mark up . . .). These two mapping strategies, neither of which is recommended, define the ends of a spectrum of possibilities. Simple mapping leads to chaos, while complex mapping is too unwieldy and time-consuming. Depending on the project, actions permitted on components will vary. Each group must think through its own relationships, balancing its strategy based on requirements, schedule, and the size and distribution of the project team.

Consider the Website description in the push-pull discussion. The three Website categories—private, public, and restricted—indicate three fixed levels of access privileges to information. The team assigns documents or other components to one of the three access levels. While this simple approach is adequate in many situations, imagine what will happen if the version-control feature team wants to take one or more of the following actions:

- It wants to restrict specific documents from all but one of the other feature teams.

- It wants to grant revision privileges for a specific document to one individual not on the feature team.

- It wants to grant the authority to approve different documents to different team members.

As with establishing nodes and links, building relationships between people and components can be simple or complex depending upon the project. Each member needs the information necessary to do the job. In fact, to work and survive at the edge of chaos, each needs *more* than necessary information, but not such an overwhelming volume that the whole team sinks into chaos. There is a delicate balance at the edge, and balancing the amount of information flowing through the team network is part of the challenge; the goal is to provide enough information to produce effective, innovative results while reducing information clutter. Another part of the solution to these collaboration issues involves sophisticated tools.

Collaboration Tools

Managing distributed development resources requires the integration of practices and tools that are emerging from a variety of sources. Unfortunately, much of the effort today is focused on using new technology from within the old process-improvement and workflow mindset. Collaboration (or groupware) tools, project management tools, e-mail, on-line discussion groups, and configuration management are all technologies often thrown together without a common integrating theme. What is needed is a *collaboration service layer*, a combination of technology and human facilitation, which defines a framework of people, a network of product components, and the interactions of people and components that are necessary to successfully build products.

This layer, which in today's world would be implemented through Web-based technologies, would provide automated support for the deliverable components and team networks. This collaboration structure would not replace existing tools (such as configuration-management, CASE, or project-management tools), but it *would* provide a context and integration layer to these other applications. The collaboration service layer should also provide state-management support for the Advanced Adaptive Life Cycle.

Turbulence and rapid evolution define the market for these software tools. In the past, many of the groupware tools have provided minimum, if any, support for product components. Conversely, component tools such as configuration control were not linked effectively to team members. However, as this book goes to press, vendors are rushing to incorporate missing pieces in their offerings. Software-development, project-management, knowledge-management, and collaboration tool vendors are rapidly forming alliances to create integrated collaboration support services.

A factor contributing to adaptive project success is shared work space—a war room or team meeting place. This physical collocation usually—but, as we all know, not always—enhances collaboration. As project teams grow larger and more dispersed, generating creative collaboration requires an equivalent virtual shared work space—a cyberspace team room. Adaptive teams need a team-owned place in cyberspace where the team can share context and content, where team members can interact one-on-one or in groups, where information can be both public and private, where there is an element of both work and play—a comfortable site to visit and to use.

The range of collaboration tools has expanded greatly during the past few years. While it is outside the scope of this book to examine many tools in detail, I include some brief examples to illustrate how the tools could be utilized to extend the reach of various collaborative practices. My list is compiled from a random sampling of tools and is not meant as an endorsement of any specific product, although I do provide product names in several examples.

Electronic Mail. E-mail systems are pervasive in most corporations today. They are used successfully for a wide variety of communications, but are more problematic as collaboration tools.

Group Calendaring and Scheduling. Individual and group calendaring and scheduling products can be used to schedule in-person and real-time virtual meetings.

Asynchronous Data Conferencing. One form of asynchronous data conferencing is the threaded discussion group. Particularly when facilitated, these discussion groups provide more structure than e-mail messages filed in topic folders. Discussion groups can be useful for dealing with ongoing topics in a development project. For example, product architecture issues are usually discussed throughout a product's evolution. In addition to the latest architecture documentation, discussion threads provide context as to how architectural decisions were made and alternatives that were considered and rejected.

Synchronous Data Conferencing. The difference between synchronous and asynchronous data conferencing is somewhat analogous to the difference between communicating via a telephone conference call and through exchange of a series of telephone answering-machine messages. At some point, serial messages just don't converge on solutions quickly enough. Similarly, requirements JAD sessions enable a variety of people to interact with each other, instead of having an interviewer serially gather information from them. For example, Microsoft's NetMeeting provides conferencing facilities for audio, video, chat, white-board viewing, application sharing, and file transfer.

Electronic Meeting Systems. Electronic meeting systems (EMSs) offer benefits to both collocated and distributed groups. EMSs can support decision-making and problem-solving meetings. An EMS might be useful in helping a group investigate and decide upon quality criteria for a product.

Facilitate.com, a product from Facilitate.com of San Francisco, is an EMS product. The meeting database can be used to list names of participants, meeting objectives, agendas, and background information. Participants can access the information using a Web browser. The tool supports meeting processes such as data gathering, brainstorming, and voting. As might be expected from the name, the product is oriented toward the use of an on-line facilitator to keep the meeting on track.

One of the advantages of an on-line tool is that while some participants are discussing basic definitions of an item, others can be drilling down to view more in-depth information about it. A lot of information can be generated quickly. When the new input begins to wane, the facilitator can then "flip" the meeting to a voting mode. The facilitator creates a ballot and distributes it to the participants, each person ranks his or her choices, and then the facilitator closes out the voting. Because the software makes the results immediately available to the participants, the group can quickly tell where there is a degree of consensus and where people differ. The meeting can then go through another round of discussion and voting.

While lacking the personal interaction of a face-to-face meeting, the on-line meeting enables people to participate who may be constrained by issues of time and travel cost. Other benefits include rapid gathering of data, data reorganization and viewing options, anonymity (depending on the meeting design), encouragement of wide participation, and moderation of disruptive personalities.

Combination Tools. There are some new tools that are beginning to offer a fuller range of collaboration capabilities. Two interesting ones are StarTeam from StarBase Corporation, based in Santa Ana, California, and Project Community from Neometron, the Redwood City, California, research and development company founded by Adele Goldberg. While these tools provide a combination of features, neither directly supports the state information required for Advanced Adaptive Life Cycle management.

First, StarBase's tool is of interest because, from its origins in configuration management, StarTeam has expanded into a product with features integrating source-document creation tools (for use in creating documents and data models), configuration-management tools, and a threaded-discussion-group tool through which topics and messages can be cross-referenced to any source document. Using a Web interface, StarTeam provides considerable flexibility in establishing and maintaining relationships between teams, individuals, and deliverable components. StarBase builds collaboration on top of Software Configuration Management (SCM).

Second, on my list of notable combination tools, Neometron's Project Community combines aspects of project management, collaboration, and knowledge management, focusing on the critical decision-making aspects of project management and providing specific tools for reusing and adapting project management and collaboration practices. It also contains "Project Model" and "Community Settings." While each project community is unique, the "Project Model" template provides a collection of best practices that may be adapted and reused by multiple communities. "Community Settings" is used to produce specific deliverables, and can identify reference information needed for an activity, define process steps, establish collaboration practices, and create monitoring data.

Collaboration and knowledge-management tools are rapidly evolving as we enter the twenty-first century. They can enable better structural collaboration, but they won't provide the right context and content unless careful thought is given to a project team's needs.

The Collaboration Facilitator

Earlier chapters discussed the role of a facilitator in JAD sessions, customer focus groups, and software inspections as someone who orchestrates the group and helps it run smoothly and effectively. Since every book on holding effective face-to-face meetings stresses the importance of the facilitator role and since virtual teams, in order to jell, need more guidance than collocated teams, the need for a facilitator should be obvious. In fact, many IT organizations have entire groups of trained facilitators who participate in a wide range of meetings, although technical JAD sessions form the bulk of their work. Skills acquired from JAD sessions form a basis for an expanded role—from JAD facilitator to collaboration facilitator. The change in title reflects expanding the responsibility from facilitating JAD sessions to assisting in a wider array of group activities and to hosting meetings in cyberspace—virtual team gatherings.

One reason why a collaboration facilitator is so valuable to a product development team has historical roots. Many IT groups have been using meeting facilitators extensively for years, while package software vendors have used them infrequently. Software vendors are sometimes so singularly focused on the contributions of individuals that their meeting skills are less effective than those of IT groups. Their typical response to ineffective meetings is either to limit the number of meetings or to refuse to attend; only rarely do such organizations try to turn the time into more productive time. Some software companies have a reputation for outright hostility even in response to personal visits. Anyone dropping by to discuss product issues is summarily dismissed with a "send me an e-mail on that." While uninterrupted time for any individual is clearly at a premium, total disregard for the rest of the team is not the right answer either.

A collaboration facilitator, then, in conjunction with the project manager, can provide a range of services to the project team over and above the traditional in-person meeting facilitation duties. He or she can

- set up and maintain the collaboration network structure

- administer on-line meetings (including plans, agendas, invitations, schedules, or discussion summary reports)

- facilitate on-line meetings (encourage full participation, assist in the decision-making process, keep the discussion on the topic)

- moderate on-line discussion forums and edit material to reduce clutter

- create and maintain effective filters to manage the information flow related to the team's development efforts

- arrange face-to-face meetings to enhance the interpersonal side of *context*, helping the project manager to balance both the in-person and the on-line mix

- make sure all participants have the appropriate tools and the know-how to use them

- understand the range of collaboration tools available to the group and help determine which tool is most appropriate for each type of meeting

A collaboration facilitator helps maintain a useful balance between content information and context information. As team networks get larger and more dispersed, the addition of skilled professional facilitators can have a significant impact on the effectiveness of collaborative efforts.

Collaboration and Emergence

Emergence may appear to be an elusive concept, but since it is at the crux of adaptive development, it is worth addressing in more depth in the context of this chapter's subject matter. Emergence lies somewhere on the continuum between axiomatic (that is, something related to a rule, a principle, or a law) and accidental.

> **Emergence is a property of complex adaptive systems that creates some greater property of the whole (system behavior) from the interaction of the parts (agent behavior). This emergent system behavior cannot be fully explained from the measured behaviors of the agents.**

The concept of emergence raises some interesting questions—for example, "What is the difference between an emergent result and an accidental one?" and "Is emergence recognizable and actionable, even if not fully explainable?"

One example of an emergent property is memory. Memory is a collective property of neurons, but how memory arises from its biological and chemical parts is one of the great mysteries of science. Do we discount memory because the rational, scientific, cause-and-effect explanation is not there? Or do we use the concept of memory because, even though it is not fully explainable, it is useful?

From a scientific standpoint, is the cause of memory unknown or unknowable? While this is an intriguing philosophical question, it probably is not relevant here. The real question is whether the concept is usable or actionable. Would anyone really argue that memory is an *accidental* property of neurons? Even if we don't have an axiomatic cause-and-effect relationship, we have a useful and actionable result. In *Emergence: From Chaos to Order* (Holland98), John Holland uses the term "recognizable features and patterns" to distinguish emergence from the accidental on the one hand, and the axiomatic on the other. We may not understand exactly how the higher-level system behavior emerged, but it has recognizable patterns.

Managers will perceive additional issues. Few, if any, of our management models have the mathematical or even experimental basis of scientific models. They have too many inputs, and the *equations* are not as provable as, for example, the physics equation $E=mc^2$. Our management models are judged more on whether they produce reasonable results than on whether they can be *proven* true. If they are usable and actionable, follow some reasonably understandable *pattern*, and can be derived from a reasonable linkage of ideas, we keep them. We continually must ask whether these models help achieve some stated mission or objective.

One important characteristic of complex adaptive systems is their perpetual novelty. Board games, like checkers or chess, generate perpetual novelty from a few rules. There is a difference between perpetual novelty and chaotic results. *Chaotic* means random, and therefore patternless. Perpetual novelty contains patterns—not axiomatic rules or algorithms—but discernible patterns that help us deal with complexity. A good checker player wins consistently by recognizing certain game patterns, even though the likelihood of seeing the exact same board configuration twice is minuscule.

The SEI's Capability Maturity Model is an optimization model whose not-so-subtle message is that anyone not in step with the SEI is immature, and therefore irrational, and therefore irrelevant. But even process-optimization models are not axiomatic—they recognize patterns. They have been successful because they work, are actionable, provide results, and follow an understandable pattern, not because they are scientifically rational and statistically provable. At absolute zero, the laws of chemistry and physics go a little haywire (for example, by exhibiting superconductivity). In complex software development environments, process optimization behaves like chemistry at absolute zero—it goes a little haywire. When dealing with any model, we have to understand its limits. Optimization models define one pattern in a vast range of possibilities, and adaptive models define another pattern.

There is no scientific *proof* that emergence is a characteristic of organizational systems. However, there is growing evidence that emergence is a characteristic that helps explain observed organizational behaviors. It is usable and actionable, and it helps organizations achieve their stated missions. The behaviors are not accidental—they follow an understandable pattern.

The Boundaries of Self-Organization

The study of emergence pushes us further into the intricacies of virtual teams and networks of teams within the context of self-organization. Although self-organization is a key to surviving in high-change environments, it is important that we know its limits.

In *The Unshackled Organization* (Goldstein94), Jeffrey Goldstein points out how some theorists and practitioners have misinterpreted certain aspects of chaos theory and its applicability to organizations. They believe that random introduction of chaotic conditions within an organization will automatically lead to self-organization and better performance. Goldstein's contention is that, to be effective rather than

"[B]oundaries define the identity of a system which enables it to explore its own creative resources when the environment challenges it. . . . I'm advocating that firming boundaries must take place contiguously with work on . . . [c]hallenging the system into a far-from-equilibrium condition. . . ."
—J. Goldstein [1994], pp. 113, 115.

chaotic, self-organization depends on firm but permeable boundaries—firm so that teams understand their part of the mission, but also permeable to the flow of information.

Without boundaries, teams lose their sense of purpose and spin off in unproductive directions. Without permeable boundaries, teams lose access to the eclectic information that they need in order to explore unique possibilities with their boundaries. *Firm* does not mean *fixed*. In fact, the high rate of information flow between nodes in the network, seen in contrast to the organizationally filtered information flow in traditional hierarchical organizations, helps to keep the boundaries from solidifying into brick walls.

There are two kinds of boundaries—organizational and product—that are reflected in the structure of the team network and of the components in the product being developed. The components of a mission define an overall boundary for a product development effort. The development team has considerable leeway within those boundaries, particularly in how it achieves the mission, but even the mission itself is somewhat ambiguous in extreme environments. While the product is expected to match the broad vision, the details are open for continual reevaluation.

Self-organization is not a call for every team to start at ground zero and contemplate its own reason for existence. Teams need to understand their roles and responsibilities, what part of the mission components they are responsible for producing, and how they need to interact across team boundaries to produce those components.

Order for Free

The study of scientific phenomena has historically involved two intertwined processes—theory postulation and direct experimentation. Nobel prizewinner Murray Gell-Mann postulated the existence of quarks, subatomic particles that provide a framework for the hundreds of particles physicists had identified in the early 1950's. Several years and many experiments later, faint traces on photographic plates began to confirm Gell-Mann's theories. Sometimes, experiments are designed to prove theories; sometimes, experimental results give rise to new theories. Unfortunately, experimenting with phenomena that involve vast possibilities—evolution, for example—requires vast amounts of time.

It is only recently that scientists have begun to view computer simulations as a third possibility; but even now, many theoretical and experimental scientists consider simulations a poor substitute for *real* science. Undaunted, Stuart Kauffman and his colleagues at the Santa Fe Institute have been working under the hypothesis that carefully constructed computer simulations can provide an intermediate level of knowledge between theory and direct experimentation.

In *At Home in the Universe: The Search for the Laws of Self-Organization and Complexity* (Kauffman95), Kauffman describes his development of Boolean network simulations (which he calls genetic networks) to study evolution, complexity, self-organization, and spontaneous order. While establishing the link between these simulations and organizational collaborative networks is in its infancy, the potential for what we can learn from Kauffman about complex human self-organizing networks is too great to ignore.

Kauffman's basic premise is that natural selection was necessary to, but not in itself *sufficient* to have caused, the evolution of complex organisms. Natural selection, by itself, faces awesome mathematical odds. In his genetic simulations, the potential variations are hyper-vast—$10^{30,000}$. However, in certain situations, Kauffman found that the model "would settle down and cycle through a tiny, tiny state cycle with a mere 317 states on it. Order for free" (Kauffman95, p. 100). The situations and conditions described by Kauffman define the edge of chaos where useful patterns emerge from vast possibilities.

Self-organization arises when individual, independent agents (cells in a body, species in an ecosystem, developers on a feature team) cooperate to create emergent outcomes. The tendency is to view this phenomenon of collective emergence as accidental, or at least as unruly and undependable. Research into self-organization is proving that view to be wrong.

Kauffman's simulations help raise fundamental questions about collaboration. Is an organization's ability to adapt quickly a direct function of its ability to collaborate? And, more pointedly, is that ability to adapt a function of an organization's collaborative structure?

Change-management strategies and change agents, brought in to help organizations change their culture, have gained immense popularity in recent years, as companies try to keep up with turbulent markets. Small companies in rapidly changing markets seem to change fairly rapidly, without much ado about formal change processes. The more interesting phenomenon occurs when some larger organizations that operate in rugged market landscapes exhibit adaptability without instituting extensive formal change-process efforts. Why does their kind of change happen? A first response may be that the market demands nimbleness if an organization is to survive. But that response does not explain *why* certain companies are better at adapting than others.

One inference I can draw from theory, simulations, and experience with organizations is,

> **The *structure* of an organization's collaboration network has significant impact on its ability to produce emergent results, and ultimately on its very ability to adapt.**

Companies establish *skunk works,* formed as separate organizations outside the normal hierarchy, to foster innovation when the structure of the larger organization is too rigid. Look, for example, at Netscape—it doesn't have a skunk works; it *is* one. Could it be that much of the effort applied to change processes in larger organizations is wasted because their collaboration *structure* is not conducive to adaptation? Should these firms be spending at least as much time experimenting with their collaboration network structure as with the latest change-management technique? Many firms spend millions trying to promote change, while their collaboration structure is solidly locked into stability.

Tuning Collaboration Networks

"Our intuitions about the requirements for order have, I contend, been wrong for millennia. We do not need careful construction; we do not require crafting. We require only that extremely complex webs of interacting elements are sparsely coupled."
—S. Kauffman [1995], p. 84.

Based on my reading in the field of complexity research, my experiences with software development teams, and Kauffman's work, I have several observations to make about tuning collaboration networks. First, it appears that above a certain number of connections, the degree of adaptation decreases. While the absolute number of connections is the key parameter in Boolean networks, the *organizational* aspect of a connection is key here. Organizational connections would include characteristics such as flow, diversity, and richness of the connection. Complexity studies indicate that adaptation occurs at the *edge of chaos,* the transition zone between confusion and stagnation. Kauffman's work helps validate this edge-of-chaos concept, and, in addition, shows that the edge appears to be a fairly sharp one. The optimal number of connections needed in order to maintain *balance* appears to be relatively small.

The optimal, and relatively low, number of connections per node (in this instance, considered to be people or groups that define a unique destination in a communications network) does not seem to vary much with network size. As networks get larger, and more nodes are added, the number of connections to each node must remain relatively constant. In today's networked companies in which a "hook'em up, link'em up, ride'em out" mentality prevails, we may be pushing ourselves unwittingly into the chaotic zone, where emergent results are lost in the noise.

Kauffman's work sheds light on how to establish the decision-making rules-base within a node. If our organizational goal is to balance at the emergent edge, we should know that overly rigorous rules override the benefits conveyed by connectivity. Although we know from experience that rigorous rules stifle creativity, we do need a better understanding of how rules and connections interact to keep organizations poised at the edge.

Finally, Kauffman establishes that there is a clear link between organisms (either organizations or animals) and the ruggedness of the competitive landscapes in which they adapt. As markets stabilize and companies settle into the ordered realm, they move away from the edge and become resistant to change. When the market changes suddenly, they cannot catch up—their control hierarchies have ousted their collaboration networks in the power struggle. Bringing in the local culture-change guru will not be enough to overcome their structural deficiencies.

Although it is a major influence in the field, Kauffman's work with genetic networks is only one of many simulations being done on complex adaptive systems. The goal of John Holland's model, named Echo, is to build a *generic* complex adaptive system model that simulates adaptation and emergence. Chris Langton and others at the Santa Fe Institute have developed another generic model, called SWARM, which they provide as a tool to other researchers. SWARM is also being used to model the richer, dynamic aspects of organizations and the knowledge that enables them to adapt. As these tools help scientists delve into the mysteries of complex biological and social ecosystems, they offer new, although still not fully tested, prospects for understanding organizational ecosystems.

"We find two themes: First, the emergence of profound spontaneous order. Second, a bold hypothesis that the target of selection is a characteristic type of adaptive system poised between order and chaos."
—S. Kauffman [1993], p. xvi.

Why Optimization Stifles Emergence

It is a widely accepted management axiom that strict, detailed procedures and bureaucratic rules impede innovation and creativity. When the topics of creativity and innovation are addressed, the discussion usually presents the perspective of individuals. Individuals feel constrained, restricted, or restrained by the organization. Software developers traditionally rebel at actual or perceived restrictions on creative freedom.

Discussions of the issue of freedom versus rigor usually focus on the needs of organizations versus those of individuals. However, what is required is more information about the environmental characteristics that support or impede *group* innovation. Seen in the context of complex adaptive systems, that kind of environmental analysis may give rise to some new ideas about group innovation .

The essential characteristic to understand about emergent, innovative results is that they arise from the interaction of many interconnected individuals.

> **Optimization stifles emergence, not only because individuals feel restricted but also because *optimization reduces the breadth and scope of interconnections and relationships.***

If we examine two interconnected processes executed by two groups, we see that process optimization reduces both the amount of information to be transferred and the number of *nonessential* connections to other groups. The more prescribed the process, the more prescribed the inputs—only the required data are communicated; communication of anything else supposedly reduces the process's effectiveness. Similarly, if the input requirements are well-known, there is less need to receive extraneous information not previously identified. And, since the context of the information is supplied by procedures manuals, there is less need for interaction with upstream or downstream groups. The end point of optimization is extreme efficiency at doing one thing—no extraneous inputs or deviations from prescribed steps, no extraneous contact with other groups.

Emergence, however, thrives on multiple connections (and the diversity they provide), rich information, and collaborative relationships. Optimizing rules operate on known inputs and known data. Emergence operates on patterns of information, and it is rarely clear what additional piece of information will complete the pattern. It is often faster to implement an optimizing process than an adaptive one because connections and relationships take time. Once organizations start down the path of "if some optimization is good, more is better," they impair their ability to establish those rich connections and relationships.

So, while some rigorous, optimizing processes are necessary as projects become larger, they should be implemented with great caution, and with consideration for their effect on relationships and information flow. The detrimental effect optimizing processes have on emergent environments is the prime reason they should be minimized, with only enough to keep the project team out of chaos and balanced on the edge.

Eight Guidelines for Applying Rigor to Project Work

One of the premises of this book is that rigor should be viewed as a balancing force, not as a goal. The discussion in this and the preceding chapter applies primarily to increasingly distributed, increasingly larger projects. In applying the practices to their adaptive project work, team members need to remember that they should not attempt to control chaos, but rather to contain it. The following eight guidelines suggest how much rigor, and of what type, to apply. The objective of these guidelines—for both management and team members—is to maintain a keen balance, right at the edge.

1. **Increase rigor slightly less than just enough.**

Rigor is a balancing force, not an end goal. Although it is necessary, rigor may stifle emergence. Procedures designed to impose rigor should err on the side of less rather than more.

2. **Increase rigor on final, integrated product components first and then move backward.**

When software itself is the final product, deliverables consist of the code base plus associated user documentation and support services. If we consider this second guideline, we know that we should apply rigor to the code-and-build components first. If this is enough to "control" the project—stop. For collocated teams, attention to these components is usually enough rigor to manage small to medium-size projects. For these projects, earlier activities in a life cycle may not require a significant degree of rigor. This does not mean that the requirements specification activity should be ignored, but that the formality and level of detail should be carefully considered.

As project size and complexity increase, the rigor applied to activities earlier in the life cycle should be gradually increased.

3. Increase rigor on components from the outermost boundary crossings first, and then work inward.

Boundaries are dividing lines between groups of people. Between companies, these are legal boundaries; inside companies, they are organizational structures—divisions, departments, project teams, or work groups, for example. The outermost organizational boundary usually divides one company from another. Outside suppliers (or customers) have the least contextual information about the product being developed, making both communication and collaboration difficult. Inside a company but outside a specific project is another significant boundary to be crossed. The boundary exists because other teams have different priorities, different schedules, and different missions, and may even compete for resources and recognition. Yet, because the two products may have common components or because one may be a supplier to the other, information needs to flow across the boundary.

Within a project team, context maintained by proximity and informality can prevail. As the distance widens, more rigor is needed to maintain the required level of context.

4. Increase the rigor of context information in proportion to the rigor of the content.

Developing documentation (content) consumes time. Often, in the race to increase the rigor (formality) of documentation, project teams fail to think about improving contextual information. Without sufficient contextual information, the additional volume of content is overwhelming rather than informing. An example of poor context would be a large, well-documented set of requirements without an index, glossary, or revision information. Asking someone to review a revised 150-page document without marking the revisions wastes time and delivers poor results.

5. Increase the rigor of context information as boundary crossings become more formal.

Within most organizations, a useful indicator of boundary formality is the number of management levels traversed to find a common manager for two groups. Two feature teams, for example, who have the same project manager can interact more informally than two feature teams who each report to a different project, department, and division manager before arriving at a manager who has responsibility for both feature teams. Although hierarchical boundaries are less rigid than in the past, they are still a fact of organizational life. As these boundaries become more formal, the need for more rigorous context information increases also.

Legal boundaries are more formal than internal company ones. Increasing the information content given to outside suppliers without providing them with context about your needs, your mission, or related details may be only marginally useful. Collaboration is more about context than content, and it is as much about personal relationships and interactions as it is about information. If there is no trust between a vendor and a project team, no amount of rigorous content documentation will provide anything more than fodder for a legal suit.

6. Increase the rigor of component information in proportion to the number of boundary crossings.

Requirements specification documents and design documents are two common development components. Requirements specifications typically are used by the developing feature team and the user documentation team, and by product support, marketing, language localization, and other departments. Many of these groups may also contribute to the specification. Each time a document is sent out or returned with revisions, a boundary crossing has occurred. Design documents are exchanged usually within a narrower audience—the development team, the testing team, and possibly other developers in groups responsible for coordinating the overall design architecture. According to this sixth guideline, a requirements document should, therefore, be more rigorous than a design document. (Note: This guideline needs to be evaluated in conjunction with Guideline 2, since they conflict somewhat in the case of requirements components.)

7. Increase rigor to support collaboration, but don't increase rigor for project management purposes alone.

If two working groups have agreed to an exchange of documents and to the degree of rigor needed for their collaboration, management shouldn't impose additional rigor purely to facilitate its monitoring of the project's status or progress.

8. **Be prepared to break any guideline.**

Guidelines are not rules. Sometimes, they will contradict each other; sometimes, they will apply; sometimes, they will be wrong. Guidelines steer judgment; they are not a substitute for it.

Summary

➤ The challenge of large, complex, fast-moving projects is maintaining some sense of the dedicated, collocated team of five-to-ten people as the number of team members grows into the hundreds and they no longer work in close proximity.

➤ In traditional projects, increasing complexity and size are managed by increasing process rigor. In adaptive projects, increasing complexity and size are managed by increasing the rigor applied to information flows.

➤ A collaboration structure is formed by interrelating the team network and the component structure. It facilitates collaboration by creating and maintaining the appropriate relationships between feature teams and the product components they create.

➤ The collaboration structure must ultimately provide tools to support self-organization. If the tools revert to traditional hierarchical control mechanisms, the innovative and creative goals of generating self-organizing systems will be lost.

➤ Virtual teams work across space, time, and organizational boundaries.

➤ The people network should consist of (1) a series of nodes connected by communications links, (2) a procedure for maintaining the network, (3) a combination of push/pull information-sharing, and (4) the linkage of people and development components—the who and what.

➤ On larger, virtual teams, the mapping of team nodes and development components is key to systematically increasing the rigor of information flows.

➤ The *collaboration service layer* is an integrated layer of collaboration tools to support adaptive development.

➤ If person-to-person meetings need a facilitator to improve effectiveness—and they do—then virtual meetings need one even more. Virtual team facilitators have a series of additional responsibilities. They graduate from being meeting facilitators to being collaboration facilitators, a more comprehensive role.

➤ Emergence is a property of complex adaptive systems that creates some greater property of the whole (system behavior) from the interaction of the parts (agent behavior). This emergent system behavior cannot be fully explained by the measured behaviors of the agents.

➤ There is no scientific proof that emergence is a characteristic of organizational systems. However, there is growing evidence that emergence helps explain observed organizational behaviors and contributes to the creation of organizations that can thrive in complex environments.

➤ The *structure* of an organization's collaboration network has significant impact on the organization's ability to produce emergent results, and ultimately on its very ability to adapt.

➤ Tuning collaboration networks (that is, instituting effective designs) through the use of complex adaptive systems concepts and simulation models may advance our understanding of the advantages of networked organizations.

➤ Optimization stifles emergence, not only because individuals feel restricted but also because optimization reduces the breadth and scope of interconnections and relationships.

➤ As speed and change increase, rigor needs to increase *proportionally,* but it must not stifle creativity and innovation. The eight guidelines help keep the proper proportions.

➤ The most important message of this chapter is that creating and maintaining a large, distributed collaborative network requires significant thought and effort—but it is ultimately rewarding. It improves both the results and the quality of work life—not a bad combination.

CHAPTER 11
Managing Project Time Cycles

Adaptive projects don't just happen; in order to be successful, they demand skilled project management. While the focus of project management in an adaptive environment is on leadership, collaboration, and accountability for results rather than on detailed task-level control, the key organizational and deliverable management activities that make projects successful must still be well executed. The project manager needs a framework in which to make critical decisions about both optimizing and adapting practices. The project management framework shown in Fig. 11.1—a generic framework used by a large number of organizations—provides a way of organizing and integrating many of the principles and practices covered earlier in this book.

Skilled project management can't accomplish the impossible. Poor management will surely extend the project's schedule just as better management will reduce it, but neither the project management framework presented nor the general adaptive approach to development is geared to accomplish the impossible or the irrational. Speed is achieved by adroit adaptive management, not by management's irrationally setting due dates.

Accelerated time frames force the project team to deal with issues quickly and efficiently. In extreme projects, time delays can prove costly. A time-oriented project management framework helps organize all the practices a team needs to survive in an extreme environment.

Operating at the edge of chaos might be interpreted as reason enough to abandon *any* traditional management practices. However, as the previous discussion on balancing rigor and flexibility affirmed, there is always a need for a prescribed set of traditional practices to maintain the potentially stable portion of the environment.

This chapter does not provide an in-depth view of project management, but follows the four-phase model named in Fig. 11.1 to address project management topics that are particularly relevant to guiding adaptive projects. The chapter provides a framework for using material previously covered (for example, product mission profiles), additional detail on relevant material (for example, time-boxing), and additional project management guidance (such as, how to contain change).

A Project Management Model

As shown in Fig. 11.1, the project management model names four key phases—initiate, plan, manage, and close.[1] Although depicted as serial steps, activities in these phases are performed iteratively in each development cycle. Activities in the Manage phase, however, are performed continuously throughout the project. For example, at the end of each cycle, customer focus-group change requests are one of the inputs to a *replanning* activity. Postmortems, one of the closing activities, may be done, in abbreviated form, at the end of each cycle.

Many project managers make the mistake of thinking project management and development activities are part of a single methodology. Some proponents of specific software development methods such as object technology, for example, believe there is such a thing as, say, object-oriented project management! I maintain that software development and project management practices are, and should be, distinct. To my mind, some software development life cycles mix the proverbial "apples and oranges." For example, estimating the project's size—a project management activity—is different from but just as important as defining the requirements of a feature. Many traditional commercial methodologies also make the mistake of integrating development and

---

[1] Although the names for these project management phases are quite general, the ideas for this model and for some of the supporting detail-level steps are derived from material developed by my colleague Lynne Nix.

management activities. This integration is an attempt to *simplify* the process—but it merely assures that every project is shoehorned into a fixed process. Each project needs an integrated but separate software development and project management life cycle.

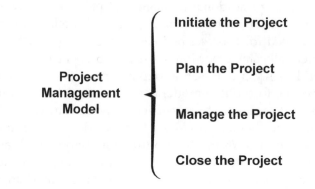

Project Management Model

Initiate the Project

Plan the Project

Manage the Project

Close the Project

Figure 11.1: Phases of the Project Management Model.

Adaptive project management is built on the framework of the Leadership–Collaboration model. So, while project management activities are similar to those in a traditional project, the conceptual framework often alters their implementation.

Let me reiterate one point before we move on—project management is a team effort. The day is past when a project manager could hide in his or her office for two days, or two weeks, and emerge with a viable plan (whether this was ever the case is debatable). Collaboration doesn't begin by having a project manager say, "Do this." It starts by having all the project team members say, "This is what we need to do."

Initiate the Project

Preparation is key to a successful mountaineering ascent. Preparation may seem to be the easiest task technically, but it is one of the more difficult to accomplish in reality because it is the unglamorous, mundane work for which few have enthusiasm. In climbing, preparation requires logistics, travel planning, and conditioning—a knowledge of what to bring, how to get it there, and how to endure the physical challenge. A weekend climb within easy driving distance requires a relatively low level of preparation, whereas preparation for a major expedition to, say, Mount Everest or K2 may require years. In software development projects, preparation is embodied in project initiation and planning.

In planning a climb, many decisions must be made that trade off safety, pack weight, and comfort. Camping gear, climbing gear, footwear, clothes, first-aid supplies, and food must all be carefully considered. Individuals spend long hours getting in top physical condition. I, for example, used to haul a backpack filled with 65 pounds of rocks up and down a long flight of concrete stairs beside our driveway, much to the amusement of my wife, daughters, and cats! Whatever the situation, it takes a lot of time and thought to make these preparations. Good preparation does not overcome the technical difficulties in the mountains, but inadequate preparation can ruin a climb before the fun even starts.

Software projects are no different. Good preparation does not ensure success, but poor preparation usually leads to failure. There are no silver bullets here. Good preparation means having the discipline to do what you must—develop a plan that includes deliverables, cycles, schedules, and the like; define and match skills to roles and responsibilities; make tool and technique decisions; develop project and product documentation procedures; choose project-measurement and change-control procedures; train; and do other miscellaneous, seemingly mundane, but critical tasks.

The purpose of the Initiate phase is to clearly identify what is to be done and under what conditions it is to be accomplished. Specifically, this phase answers the following questions:

- Who will plan, manage, and perform the project?

- What is the purpose of the project (the objectives)?

- What are the project constraints?

- How complete is our understanding of the project?

- How will planning and management be approached?

In order to plan any project, traditional or otherwise, team members must define the basic specifications (the feature set) for the application. To launch an adaptive project quickly, however, project managers must simultaneously specify that the project initiation deliverables (for example, the project data sheet) and the outline specifications be developed during an initial JAD session.

"The seeds of major software disasters are usually sown in the first three months of commencing the software project."
—C. Jones [1996], p. 51.

The project planning example given in Chapter 4 described a relatively small project, but there are points specific to initiating large projects that should be emphasized. For small projects, much, if not most, of the project initiation work can be accomplished in a single kick-off week. For larger projects, additional training, resource procurement, architectural work, and project justification may lengthen the initiation time by several months (although more than three would be excessive). If the initiation work is scheduled to take more than a couple of weeks, it should be designated as a Cycle 0. This designation indicates that project activity has begun, but that no application features will be demonstrated.

This brief consideration of Cycle 0 project initiation brings up the important topic of system architecture. As other iterative development processes (the Rational Unified Process, for example) have correctly pointed out, a well-considered architecture is important, but architecture, like requirements definition or design, should not be confined to a single project cycle. While comparatively more time may be spent defining architecture in Cycle 0 than in later cycles, it should be addressed, and possibly revised, during each cycle. Over the life of the development effort, architecture will probably be more stable than, say, requirements, but it will still change. Trying to get architecture "right" in Cycle 0 is the "wrong" approach. Keep Cycle 0, and activities like architectural analysis, short, allowing a better solution to evolve over time.

Identify the Project Team

During the Initiate phase, roles and responsibilities of and relationships and communications lines between the project team and the executive staff are defined. These include all the collaboration network components.

An adaptive project team needs a single point of contact with senior management—an executive sponsor. For an internal IT project, the executive sponsor may come from the customer's organization and may be supported by a steering committee of other customer representatives. For a project done by a software vendor, the sponsor is usually an executive from one of the product-development organizations. The key responsibilities of an executive sponsor are to articulate objectives, approve resources, and establish constraints. The project team members can help develop detailed objectives but can't set business objectives; they can also utilize assigned resources but cannot alter resources assigned; and they must operate within established constraints.

An executive sponsor is important to the success of any project, but especially to an accelerated one on which clear direction and assistance are more critical. The executive sponsor must understand how an adaptive project is different from a traditional one, particularly regarding the specific life cycle, the emphasis on product components, the applicable view of progress monitoring and management accountability, and the need to assist the team in procuring resources and in quickly reaching decisions.

The core team, one for each major feature set, consists of people who are committed full-time to the project. This core team for an extreme project must be dedicated to the one project only because part-time participation (which hinders any project) may totally derail high-speed, high-change projects. While core-team members drawn from the customer's organization preferably should be willing to participate full-time, in many cases, the reality of the situation is that the "right" people are just not available full-time. In accelerated projects, the key to success is for client team members to be available to the team on a daily basis. Having the "right" person available every day, even for only two-to-three hours, is often preferable to having the "wrong" team member just because he or she happens to be available full-time.

As was described in Chapter 8, the project leadership role drives success. The project manager's role is a full-time effort on most adaptive projects. The speed and change of an adaptive project usually keep the project manager's job very intense.

Create the Project Mission Data

Although the task of creating project mission data was detailed in Chapter 4, one point bears additional discussion in terms of the Initiate phase. The question is, How much time should team members spend creating mission data?

For smaller projects, the mission data-creation activity might take a few days; for very large ones, it could take a month or more. Time should be spent by team members reviewing historical information about the project that has been accumulated over time by clients, marketing staff, or developers. This historical information may range from a detailed market research study in a product company, to a project proposal and cost/benefit analysis put together by an internal client, to pre-project work done on one of the IT group's projects. The primary purpose of this step is to review this information and gather any additional information (project mission, project planning, or product requirements, for example) required, in order to clearly identify what is to be done and under what constraints it is to be accomplished.

Define the Project Approach

One problem with many projects is that the team fails to discuss and agree on the approach to be used in executing the project. Roles are confused, arguments arise on the use of development techniques, tools are introduced late in the project, and other needed equipment or materials aren't obtained on time. These problems are all symptoms of the team's failure to adequately establish the framework for planning and managing the project. While these areas are detailed at the beginning of a project, they should also be considered as part of the review at the end of each cycle. During an adaptive project, decisions must be made regarding activities in each of the following five areas:

Roles and Responsibilities: Determine who will be involved and what each person's involvement will be during the planning and development phases of the project.

Communication and Collaboration: Identify all of the project's stakeholders, what information each stakeholder needs, what information each stakeholder will receive, and the means and schedule for providing the information. Reach consensus on all the initial collaboration structure decisions and on mappings among deliverable components and teams.

Development and Learning: Identify which development practices and support tools will be used during the project. Also identify which learning practices, such as customer focus groups, will be used.

Change Management: Describe the approach to be used for both change control and change containment. Identify and define the baseline items of the project and determine how changing these items will be managed during the development process.

Progress Assessment: Identify how and when project progress will be measured and assessed, how exceptions will be identified, and how adaptive actions will be planned and implemented.

Increase Speed by Starting Early

It is always amazing to me to see how much time is wasted getting projects started. I have seen months, even years, wasted, with the development team then being admonished to speed up to a completely unreasonable delivery date.

Accountants do not speculate very well. Many project budgets depend on having firm cost estimates before a project can be approved. Teams are reluctant to build *anything*, even a prototype, without significant cost/benefit justification. The effort becomes a vicious circle—no money without detailed paperwork and justification, but producing these requires a better understanding of the scope and features, which in turn requires money. Thus, many firms waste months and months because the effort to define features and scope is made into a part-time activity so it won't cost much. The dollar cost is low, but the cost in time is enormous.

I've watched several software companies spend more than a year doing feasibility studies and still end up with a very poor understanding of the project's scope. A better approach, if the project is large and the product appears to have good market potential, is to concentrate resources for two-to-four months (or fewer, if possible) first on feasibility and then on building a prototype product. Spend at least half the time building one or two cycles' worth of features. The working model will provide decision-makers with a much better idea of the product's scope and feature set than would documentation alone. The working model can also be shared with the potential customer base. This work before Cycle 0 might be thought of as Cycle –2 and Cycle –1. In less than half the time one might spend in traditional project identification (work done prior to initiation), the product can be better scoped, and the size and cost more accurately estimated. The approach requires some up-front money and the willingness to speculate, but it can pay off in significant schedule savings.

"The front end [prior to official project start] offers some of the cheapest opportunities to cut development time that are to be found anywhere in the cycle. . . . [T]he front end is so fuzzy that people tend to forget that it even occurs. . . . [W]e have seen situations where as much as 90 percent of the development cycle elapsed before the team started work."
—P. Smith and D. Reinertsen [1997], pp. 44, 46.

Plan the Project

Project initiation establishes the broad framework—mission, background data, organization, and approach. Planning fills in the details—a product size estimate, number and length of cycles, detailed resource analysis, and more. Inevitably, there is an iterative interplay between initiation and planning as more information is acquired.

The Plan phase ensures that all project team members understand what must be done and the time frames required for completion. The goals of the planning phase are realism, flexibility, schedule viability, and establishment of a collaborative environment.

The effort spent during the planning phase will directly impact how smoothly the team can implement its plans once work begins. The planning phase addresses a series of questions:

- How large is the application?

- What are the project and cycle time-boxes?

- What features and components are included in each cycle?

- What are the dependencies between components?

- Who is responsible for each component?

- Are the resources assigned to the team adequate?

- Where is the project most at risk?

- What is the least understood part of the project?

The project team needs to address each of these issues in the depth and breadth appropriate to the project. The basic activities to be carried out during the planning phase follow:

- Define the work.

- Develop the project schedule.

- Analyze the resource requirements.

- Assess project risk.

Project planners, just like mountaineers, need to fully understand the challenge ahead of them. Project planning involves more than constructing a few Gantt charts and assigning generic resources. Planning complex projects requires not only good execution of defining, estimating, and scheduling tasks, but also requires good judgment of the nonquantifiable—risk, uncertainty, teamwork, collaboration, and creativity.

Extreme projects, like mountains, share similar characteristics and problems, but no two are identical. Mountaineers, for example, instantly understand what peaks classified as eight-thousand meters will entail even though they may never have climbed the specific peak itself. They know that climbing an eight-thousand-meter peak involves significant time, money, and preparation. It also involves significant risk, to which the spring 1996 tragedy on Mount Everest, in which twelve people from several expeditions died, attests.

But even weekend recreational climbers need a classification system to keep them from getting in way over their ability level.[2] While a one-pitch, 5.10 sport climb might make a fun afternoon outing for a reasonably proficient recreational climber, a 5.9X traditional climb (for which X indicates seriously poor protection with a significant likelihood of injury if a fall occurs) is an entirely different undertaking, at least as far as risk is concerned. And a 5.11, Grade V climb indicates that a climber faces a two-to-three-day ascent and nights spent gently rocking to and fro on a portable two-foot-wide sleeping ledge suspended by ropes. The mountain climber knows something of what lies ahead by virtue of the rating. The project team member also should know what lies ahead when a project is rated extreme, and he or she must use the Plan phase to prepare to meet the challenge.

Before we look at how individual activities within the Plan phase can be structured, we should investigate the concepts of time-boxing and resource fragmentation to see how they impact planning adaptive projects.

Time-Boxing Projects

> **Time-boxing provides a structure for focusing on the most important mission characteristics and forcing engineering trade-offs.**

As the above definition suggests, time-boxing is not only about time. A client once lamented to me, "How is time-boxing any different from what we have always done? Management gives us impossible schedules, then we work until it gets done, usually way past when they wanted it." The frustration expressed in this client's statement is exactly what time-boxing is designed to prevent.

Too many developers and managers have misinterpreted time-boxing as being the inflexible servant of time. In order to deliver results in complex environments with high levels of uncertainty and conflicting

---

[2] To help readers make the most of this analogy, here are some brief definitions of special terms—*Pitch:* a distance on a steep grade measured as if by one rope length (165 feet); *5.0–5.14:* climb-difficulty grading system; *Sport Route:* a route with anchors (protection) that are permanently fixed in the rock; *Traditional Route:* a route along which each climber places removable anchors while ascending.

constraints, managers must have available some mechanism to force periodic convergence. Without some forcing function, the complexity can overwhelm the team and cause prolonged periods of *groaning*. Every time convergence is forced, the results are subjected to review and revisions. All the components of the quality equation are examined and plans are modified appropriately. Whether we like it or not, schedule slippage is a fact of life in the development of complex products—the uncertainty is just too great. The key is using time appropriately, not as a bludgeon, but as a forcing function to periodically converge on a solution.

Effective time-boxing is built on the same premise as other good project management techniques: Schedules should be aggressive, but not irrational. Irrationality can come from a variety of sources, but two having a particular impact on software projects are sizing and trade-offs. Without adequate sizing, scheduling is a shot in the dark and delivery dates are fantasy. But even with reasonable initial sizing, the evolving requirements, alternative design considerations, and other project changes require adjustments. Adaptive projects need good size estimates, but since we know the estimates won't be accurate, time-boxing is needed to force hard trade-off decisions and alterations to the project plans. Estimating is like any other part of an adaptive project—one must speculate and adapt rather than plan and control.

There is often a sense among technical staff that faster speed always means poorer quality, but the problem is more complex than a simple, linear trade-off. First, many of the adaptive techniques offer the promise of higher quality through lowered defect levels. Iterative testing, for example, can help catch serious errors earlier in the product's development. Second, adaptive development stresses the difference between quality and defects—that defects are one part of the quality equation. The software project team that delivers technically superb union-payroll modifications three months late or the developers who deliver commercial software five months after a competitor has delivered its software may indeed cite their technical quality while looking for a new job. Staff and management must always balance speed and quality; unfortunately, the trade-off is not always clear.

Time-boxed project management does not mean cutting corners to make a schedule date at the expense of other quality attributes. This misunderstanding has led many who are concerned with technical quality characteristics such as defect levels and maintainability to see time-boxing as a dangerous gimmick.

Adaptive development uses time-boxing to focus attention on the most important quality characteristics and to force engineering trade-offs. People who are most worried that time-boxing will result in inferior technical quality often have poor specification and measuring systems for those attributes, and therefore add processes and time to the development schedule in an attempt to satisfy nebulous goals. The definition of time-boxing mentions mission characteristics—a category that includes quality measures—not processes. Rather than add processes to a project in order to enforce some ill-defined view of quality, adaptive development relies on defining the quality attributes and letting those dictate which processes are necessary. For example, developing an IT application with a very low defect-level specification virtually demands that some form of inspection process be in place.

Time-boxed project management means using good application sizing and estimation techniques, such as counting function points, adjusting estimates for other quality requirements, and then using available project resources to develop a rational schedule. This overall schedule is then divided into development cycles, each with its own time milestone. By establishing fixed times, time-boxing forces trade-offs in other areas such as features and resources. It also guards against the danger many engineers face in believing in perfection and having a tendency to overdo.

A good guideline for time-boxing on various cycles of development is to maintain strict deadlines on the first two cycles and less strict deadlines on the final two. Keeping strict schedules on the first two enables team members to determine whether the initial estimates or execution were off. In this way, the team forces itself to determine the viability of the initial plan. If changes during the project have begun forcing schedule slippage, time should be added to later cycles, not to early ones.

Time pressure can be a very positive force in software development, but it is also one of the most frequent causes of failure. Using time effectively is the goal of time-boxed project management.

Staff Fragmentation

Fragmentation occurs when individuals (or teams) attempt to work on many tasks concurrently.

Staff fragmentation is one of the most severe and most misunderstood problems facing software development project teams. I am convinced that at the top of the list of reasons why adaptive projects succeed is their insistence on dedicated teams. Many organizations play games with staff resources. As projects continue, especially those lasting more than six months, there is a tendency either to reassign team members to new projects or to assign them temporarily to other work. Loss of staff to other projects lengthens the first project. As projects drag on, the pressure to *borrow* their resources increases. In an attempt to show customers that something is progressing on *their* projects, more and more projects are started, with fewer and fewer people on each.

In *Why Does Software Cost So Much?* Tom DeMarco reveals the difficulty in answering managers who ask, "What if we did only *one* thing to improve our chances of success? What would it be?" (DeMarco95, pp. 87–88). While DeMarco concludes there is no one right answer to these questions, he does point to staff fragmentation as a major contributor to project failure. He adds that far too many companies' "people are frustrated by too much task switching" (op. cit., p. 89). For example, people work on several projects, take classes, attend meetings, and serve on process-improvement teams. DeMarco contends that the occurrence of too many complete context switches per day causes thrashing, just like the thrashing that occurs in an overloaded or badly designed multi-tasking operating system. He concludes that "three complete context switches use up a whole working day" (op. cit., p. 90).

Some managers might defend task switching by arguing that they have to assign staff members to multiple projects or those people periodically would sit around doing nothing and waiting. Managers want to fill every available time slot to maintain efficiency and productivity but unfilled time can be an advantage rather than a disadvantage. Bottlenecks are not caused as much by a little idle time here and there as they are by too much work-in-process—that is, too many projects and tasks to be worked on concurrently.

Fragmentation is also caused by frequent interruptions, such as those caused by the telephone, coworkers, loud speaker announcements, and unnecessary meetings. However, fragmentation is not just the organization's responsibility. Fragmentation is both an individual and an organizational concern. We allow ourselves to be fragmented; indeed, we encourage fragmentation: We respond instantly to return phone calls or send e-mail rather than waiting until a logical break. We start off looking for something specific on the Internet, say, and end up off in another zone. We respond instantly to interruptions, possibly because to do so is more fun than the longer-term project in front of us.

Adaptive teams must balance people's need for collaboration with their need for uninterrupted working time. The more extreme and intense the project, the more focused the team must be, and therefore, the more it must resist fragmentation.

Define the Work

Most details for *planning* the adaptive cycles of the project were presented in Chapters 4 and 9. However, *defining* the work can be a challenging task. How the work for each cycle is to be defined and selected is determined by the project team first identifying all components—product, technical, and support—and then any secondary tasks that must be done to support the effort.

In some projects, creating one task for each deliverable component allows staff to use traditional project management tools. For projects with a large number of concurrent components, traditional tools need to be supplemented with collaborative services and tools similar to those referenced in Chapter 10.

For a project to be successfully completed, the work to be done must be identified and organized. In an adaptive project, the interim product cycles provide concrete deliverables with which to assess progress. Because these cycles are often weeks apart, rather than months apart (as in a traditional project), task breakdowns during cycles are not critical.

Part of defining the work involves estimating the size of the application. I recommend that function-point counting or other metrics-based techniques be used for the sizing effort. Once the size has been estimated, that and other factors determine the time-box for both the project and each development cycle.

Develop the Project Schedule

An important project scheduling issue is *target* versus *committed* delivery dates. There is often confusion and even hard feelings among developers when someone in management or marketing sets a date. The common refrain is, "How can they set a date when we don't even know what the specifications are?" Management, or marketing, often has a business need that must be met by a certain time. This date should be established as a target date, one having business justification. A committed date, however, is one the development team sets only after project initiation and planning sessions. Only then is enough information available for the team to commit to a schedule that is based on some rational data. There is then a negotiation between target and committed schedules based on solid planning information and the mission profile.

For example, staff productivity rates and project size should be used to validate the schedule as reasonable. If not, the project team might prepare alternative plans for consideration. One possible alternative would be to cut the software's functionality. A second might be to increase the staffing. Only when the project team has analyzed and planned the project is a *committed* date determined.

"The first aspect of a successful project is to utilize a formal sizing approach early in the development cycle. . . ."
—C. Jones [1996], p. 14.

Function-point analysis has been widely used to estimate the size of software products. Since the emphasis in adaptive cycle planning is on deliverable components rather than on tasks, adaptive cycle planning facilitates the use of function-point sizing. Without some reasonable sizing technique, project trade-offs become a decibel test—that is, determined by who protests the loudest. Just because the unknowns in an adaptive project may be extremely high does not mean that size estimation is fruitless. As the uncertainty of the estimate increases, techniques such as the addition of buffer time can help in managing the schedule.

Analyze the Resource Requirements

Once basic cycle planning and task planning have been accomplished and a preliminary schedule has been generated, the team must evaluate the resource requirements and answer the following questions:

- Is the preliminary staffing adequate? (Many accelerated projects have early staff assignments.)

- Has the planning effort identified the need for a different skill mix than the one currently assigned?

- Is any additional staffing warranted during the project?

- Is representation from the client organizations sufficient?

- Is additional subject-matter expertise required?

Adaptive projects are fast-paced. Obtaining resources consumes schedule time, whether the task is finding a war room where project members can convene, hiring contractors onto the staff, or getting a signed purchase order. The project team that does not identify and attack these seemingly mundane resource issues early will find itself suffering for weeks into the project, and possibly putting the schedule in jeopardy.

Assess Project Risk

Risk management is a current hot topic in software development circles. Although many books, articles, and reports have been published on the topic, Tom DeMarco has written what I think is the best assessment of the true state of risk management in most organizations.

"[T]he proactive management of risk and opportunity will become the central operational paradigm of business for the late 1990's and early 21st Century."
—R. Charette [1993], p. 6.

> *Can-do thinking makes risk management impossible. Since acknowledging real risk is defeatism, the risk management function in a can-do organization is restricted to dealing with those smallish risks that can be mitigated by quick action. That means you confront all the risks except the ones that really matter.*
> —T. DeMarco [1995], p. 214.

Risk management, according to Barry Boehm (Boehm91), is one of the most important skills of successful project managers. Risk is a measure of the probability of an unsatisfactory outcome times the loss occurring from such an outcome. Within the Adaptive Software Development framework, risk analysis can be used in two ways: first, to assist in cycle planning; second, to modify the basic plan in order to manage the risks.

Most organizations give only lip service to real risk management. It is often thought of negatively, as if making excuses for failure. One young project manager at a client organization remarked to me, "We don't need to do any risk analysis on this project; it's too important to fail!"

Risk management isn't done to make excuses; it's done to identify potentially adverse conditions and isolate them, to reduce the probability they will occur, or to reduce their impact if they do occur. To the outsider, climbing appears to be a very risky undertaking, and in certain circumstances it certainly is. But the vast majority of climbing activity is reasonably safe, because climbers understand the specific risks and make significant efforts to keep them at bay. In the mountains outside Salt Lake City, there are several instances each summer in which "climbers" are reported injured. In many if not most of those incidents, the injured were not climbers, but people without any climbing skills or training, out scrambling around on rock faces and getting themselves into trouble. They usually are the ones that have no idea of the true risk, and therefore have no inclination to manage the risk by using proper equipment and training.

Experienced climbers take risks, sometimes frightful risks, but they generally understand the situation and make an informed choice about how much risk to take. Obviously, they sometimes lose. Climbers who ignore the risks, or lose their focus on confining those risks, often pay dearly. Active risk management is an ongoing, vital part of climbing. Risk management is built into planning a climb, and into every stage of ascent. It is not an opportunity to focus on failure, but to focus on the chance of success—a constant risk/reward analysis. Just as in any mental exercise, too much consideration of the negative can hamper the result, but too little can lead one to take foolhardy chances.

Risk management plays an active role in adaptive projects. From early planning, to detailed planning of each development cycle, to the daily execution of each activity, risk management should be a part of the process. As in most things, good risk management is harder in practice than in concept. Good risk management runs afoul of politics and human frailties, probably more than most other management practices, because it demands we acknowledge our fears.

But risk management is much more than avoiding unfavorable outcomes. Robert Charette, a prominent author and practitioner of risk management, uses the term "risk entrepreneur" to describe those who use risk management as a competitive tool. Rather than manage or ignore risk, the risk entrepreneur uses uncertainty to attack it. A risk entrepreneur is similar to a world-class mountaineer: Each can enter dangerous terrain where others fear to tread because they understand the risks in the context of their own skills. They both have a greater chance of success than their competitors. Neither is foolhardy; in fact, they both comprehend the risks better than the competition does and are therefore able to exploit the situation.

Charette reminds us that with inevitable change comes the need to decide our course of action:

> *For with choice comes risk and opportunity: in other words, the opportunity to end up with something better than the status quo, or the risk to end up worse than we are currently. And given the uncertainty that most choices possess, . . . it is easy to understand why most of us are satisfied with the status quo, thank you very much.* —R. Charette [1993], p. 6.

Adaptive Software Development is a tool for the risk entrepreneur—it enables him or her to enter dangerous terrain with a greater chance of getting to the top.

Manage the Project

While material in previous chapters has focused on key aspects of managing projects, this section offers additional ideas on a manager's need to persist, monitor progress, and finish strong. Then, at the end of the section, specific pointers address containing (as opposed to controlling) change.

Persisting

During product development, and especially during the middle section of a project or cycle, project team members must be unrelentingly persistent. Tasks during this period require team members to expend consistent activity with bursts of high intensity, but they do not necessitate quite the level of sustained high intensity as is required near the end of a project, just before the product ships. Too much intensity during this middle point results in burnout and an inability to step up the pace at the end. This is also not the time to be inconsistent or sloppy, just because there appears to be enough time to ease up.

Here are some strategies to follow:

Keep communication levels high: Encourage frequent but short team meetings; post the project data sheet, risk list, focus-group change requests, and key technical models on the war-room wall (or in the virtual team room); keep the team together; encourage frequent client contact; construct effective status reports; execute the communications and collaboration plans.

Maintain focus: Periodically review the information posted; focus on developing components and avoid getting bogged down in detail; constantly refer to the product mission and profile; resist letting anyone—whether managers or clients—sidetrack the team.

Overcome obstacles quickly: Anticipate problems, make decisions quickly, and move on.

In mountaineering, especially on long glacier passages, the monotony can become overwhelming. Climbing slopes of endless snow and ice that have little landscape variation is like night-driving in west Texas, with the monotony of the snow just as dangerous as the long white line. Wearing modern crampons and using ice axes, climbers can traverse extremely steep snow and ice slopes with relative safety. However, a false step or a crampon point snagged on a pant leg can lead to a headlong, accelerated descent. In some cases, before the speed becomes too great or if the snow isn't ice-hard, the fall can be stopped by planting an ice axe. If not, the falling climber can only wait—to catapult off a cliff, fall into a crevasse, or crash into a rock outcropping. In climbing, persistence—determination to move on—overcomes the monotony of long, arduous stretches of mountain slopes; constant vigilance prevents disaster. Likewise, during the middle software-development cycles, persistent, relentless grinding out of the daily work, component by component, gets the seemingly endless job done. In risk management, vigilance wards off disaster.

Much of a leader's job during development involves hard, manual labor—removing obstacles and providing resources can be likened to moving rocks and hauling water. The late Jim Davis, a long-time colleague and friend of mine who was head of an IT organization in Atlanta, told a story about one of his most productive developers whose output had fallen off. Questioning the team leader about the developer's productivity, Jim discovered that the developer usually spent several hours working at home each evening, but now had a damaged modem. Since the developer's home equipment was his personal property, the team leader had let the situation continue, thinking there was nothing to be done. Jim, who was a tall, imposing presence, nearly exploded, "Go buy him a new modem today—hell, buy him two!" Sometimes, the rocks aren't as big as we think. Good leaders must know how to move both small and large ones.

Hauling water means getting the resources the team needs. Whether the resource is a bright new recruit or the latest PC equipment, good leaders of extreme teams must keep ahead of the resource curve. Good leaders recognize talent and recruit it with a passion. They push the boundaries of the organization's formal systems, living by the old adage, "It is easier to ask forgiveness than permission."

Monitoring Progress

The Manage phase of the project management model outlined at the chapter opening requires that the project team members and the team leader track progress and adapt their strategies to reflect new information. Questions that must be answered at the end of each adaptive cycle are

- Where is the project? (What is the status on scope, schedule, resource, and defect levels?)

- Where should the project be? (What is the deviation between goal and reality?)

- How do we adapt to get there? (What kind of alternatives do we have?)

- When will we arrive? (Can we establish a revised schedule estimate?)

In traditional, serial development, progress is measured by the completion of certain tasks. For example, after four months, the requirements analysis is *done*. Completion, even though it is ephemeral in the context of development cycles, gives a sense of closure. When using adaptive cycles, the end of Cycle 1 may find multiple components partially complete, but none finished. My experience has been that adaptive project teams *feel* least sure about progress in the beginning. In many traditional projects on the other hand, the early good feelings turn out to be unjustified. It is easy to get into trouble in traditional projects because documents give team members a false sense of security. Because adaptive projects produce parts of final products rather than whole documents, precise measurement of progress is harder than in traditional projects, but the qualitative measures are more indicative of *real* progress. As the project proceeds, adaptive team members understand much more about the final product than a traditional team does, even if they have a harder time precisely measuring the progress.

Both qualitative and quantitative progress data should be gathered for each of the four project profile components—scope, schedule, defects, and resources. In an adaptive project, scope progress is measured primarily by monitoring component completion through focus group sessions or cycle reviews.

The project team should gather enough information to answer the following questions for management:

- **Scope**
    - Did the product components demonstrated at the focus group presentation satisfy the planned cycle deliverables?
    - Do the change requests generated from the focus group seem reasonable in terms of scope boundaries?

- ◆ Do the change requests reflect a convergence of client specifications and delivered components?
- ◆ Does the potential resolution of any focus group issue appear to impact scope seriously?
- ◆ What was learned in the cycle that would significantly alter scope estimates?

- **Schedule**
  - ◆ Was the cycle completed on schedule?
  - ◆ Does the schedule for the next cycle still appear feasible?
  - ◆ What risk factors would adversely affect the schedule?

- **Defects**
  - ◆ Is the quality of the delivered components appropriate to the stage of development?
  - ◆ What are the defect levels and what are the trends?

- **Resources**
  - ◆ Have core team members been available full-time?
  - ◆ Is client availability adequate and at the planned levels?
  - ◆ Are team members comfortable with their roles?
  - ◆ Are members of the team becoming overloaded or burned out?

In answering these questions, the team must also reassess the existing project plans and develop any adaptive actions necessary to overcome problems. Differences from plan may or may not indicate a problem. Only careful analysis can determine whether a problem exists and what the adaptive or corrective actions might be. In analyzing any variations, the team members need to find out if the problems pertain to one-time, easily fixable events, or if they indicate potentially longer-term, serious problems.

During each cycle's progress assessment meeting but at a time separate from the focus group meeting, the team should conduct a mini-postmortem. This form of retrospective (described in Chapter 6) may point out changes needing incorporation into the next cycle plan.

A project status report for an adaptive project should be simple and informative, providing a snapshot of progress, problems, and corrective action plans. It should also point out areas for potential management action. Specifics about the contents of a project status report and its recipients should be identified during project initiation.

Project status reports should contain most of the following:

- an introduction, including the project-objective statement

- a recap of significant accomplishments

- specific quantitative data to indicate progress

- highlights of problem areas

- an assessment of the impact of problems on the project plan for the next cycle and for the remainder of the project

- specific corrective actions taken or anticipated

In addition to preparing periodic written status reports, the team may also need to hold cycle milestone reviews with management.

Finishing Strong

How many times have sports stars like Michael Jordan, playing for the Chicago Bulls, or John Elway, during his years as quarterback with the Denver Broncos, pulled out a last-second victory? The player's endgame is usually important. In sports, there is frequent cause to lament an athlete who has great statistics, but cannot seem to *step up* at the end. On software projects, the endgame is a time of drama, intensity, and, often, poor execution.

Redpoint climbing tests the climber at his or her maximum performance levels. The last fifteen feet of a climb can be excruciating. If the crux—the most difficult section of a climb, often lasting only one or two moves—is already past, the climber's tendency may be to relax, but to do so is sure to cause a fall on an easier section. Near the end of an exhausting task, a person's body and mind perform strange tricks. Like basketball or football or gymnastics, climbing is a sport in which fatigue and the need to maintain coordination can clash.

In climbing, fatigue and lactic-acid buildup cause the climber's brain to lose its power of concentration; as the climber's focus narrows, the body reverts to reflexive, usually incorrect, actions. If the climber had just been able to concentrate on the *right* actions, the finish might be different. The key is learning, through practice, to do the right things under high stress.

Software project teams under stress often experience a similar loss in concentration. Under stress, team members tend to go "dark" (McCarthy95), lock themselves away, and concentrate on their own tasks. Communication deteriorates. Rather than carry out practices they know they should continue, they let shortcuts become the norm. The team must *practice* working under stress so that the initial destructive reflexes don't take over. This endgame practice suits the iterative nature of adaptive development. The team that is falling apart before the first or second focus group review needs to review its performance under stress so it can finish strong in the last cycles.

Containing Change

High-change environments demand that team members understand the state of equilibrium as a *temporary* condition. To manage a project in a fast-moving environment, a leader must be skilled at both controlling and containing change. Change causes something to become different, with differences arising between what was planned and what actually occurred, or between something that was previously done and what is being done currently. Controlling change is usually perceived as a specific process for documenting, assessing, and acting on a difference (for example, revision control for a document). In traditional projects, change, particularly that arising from the difference between plan and actual, has an implied, if not explicit, negative connotation.

Containing change is less formal. Containment requires that managers and staff understand differences, decide if the differences should be acted upon, and notify other people or other teams of the actions. In adaptive projects, since change is considered the norm, differences are viewed less negatively, as items to be considered, not eliminated. Finding answers to specific questions such as the following becomes part of the project team's everyday work activity:

- What is the difference between today's partially or fully completed components and those outlined in the requirements?

- Do we have existing procedures or practices—rigorous, flexible, or problem-solving—to handle these kinds of differences?

- Whom, if anyone, do we need to notify about the differences?

- How formal must we be in recording the differences?

- Do the differences exceed an established boundary condition?

- If the differences do exceed established boundary conditions, what does the boundary crossing mean in terms of recording and notification?

Containing change requires managers to perform a kind of balancing act. In high-change environments, neither leaders nor staff have enough time to control every change with formal change control processes. To control change, one must attempt to stabilize the environment, to make sure every component is catalogued and every revision documented. Managers wishing to contain change seek to stay out of chaos, but realize that complete tracking of all changes is not practical. In an adaptive project, managers employ rigorous change control processes on some items (source code, for example), but less rigorous containment practices on other items.

Containing change is too big a job for managers alone; it must be everyone's job. Techniques for containing change include bounding, ignoring, postponing, filtering, replanning, and buffering. While these techniques are not unique to adaptive projects, they are not usually associated with the rigorous process of change control.

Bounding Change

A mission statement, a resource constraint, and the end of an adaptive cycle are all examples of boundaries that help contain change. Within a development cycle, the team members may appear uncoordinated and free-wheeling, but the cycle end funnels the energy into serendipitous results. While resource constraints may have significant leeway, monitoring them is one way of ascertaining whether the team is converging on a solution or endlessly circling. Specifying the desired results (the "what") and letting the team members decide how to achieve those results is also a form of bounding.

Ignoring Change

Many differences can just be ignored, or, at the least, not acted on at all. In collaborative meetings, for example, hundreds of differences may be discussed before a joint solution is proposed. Although selected meeting notes should be taken if they can serve as reference material to support a decision, having a meeting scribe record all the differences would be counterproductive. How much you ignore may also be a function of your tools. Newer word-processing software with advanced redlining, annotation, and versioning features makes recording and monitoring differences comparatively easy. Without these tools, "ignoring" seems to be a very acceptable option. (Having recently written a two-hundred-page document with six coauthors, I can say that keeping up with differences is not easy, even with access to advanced tools.)

Postponing Change

Sometimes, the best strategy is to postpone acting on a difference until it is time for the next revision, the next cycle, or the next product version. Postponing does involve some processing, since the information needs to be recorded so that it can be retrieved at the appropriate time.

Filtering Change

Filters, which generally contain characteristics of a component along with associated permissible actions on the component, help a manager decide what action to take on differences. Filters are particularly appropriate when cycle milestones and ship dates are approaching. In a workshop I held with a client several years ago, the participants identified a pervasive mental model they were operating under: "There is always time for one more change." In analyzing the evolution of this mental model, they realized it originally stemmed from a desire to be more responsive to their customers. However, over time, they were accepting virtually any change, from any source. Setting up filters at various time frames prior to the ship date, with criteria for incorporating changes, insured that critical customer-related changes were accepted while less important ones were postponed.

Replanning Based on Change

The end of every cycle includes a replanning activity. Looking at planned features, implemented features, new features, requested changes, and all other project components, the team adjusts the plan for the next cycle. If the adjustments violate the boundary conditions of the mission statement, additional executive sponsor approval may be necessary. As the project proceeds, the boundaries narrow, but where traditional project managers spend more time explaining and justifying variations, adaptive project managers have a tendency to accept the differences, adjust to them, and move forward. Functioning in the realm of disequilibrium and unpredictability, adaptive team members know that time spent justifying differences is not productive.

Buffering the Project Schedule

Time buffers are unallocated time periods added to the end of projects to help manage uncertainty. They should be used to react to unforeseen events and misestimates based on inadequate information. Time buffers help compensate for the fact that, although we know that changes will occur, we don't know what they will be. Some kinds of changes can be anticipated; others come out of left field. Once the time allocated for a cycle plan is established, 10 to 20 percent more time should be factored into the schedule. The wise team targets the original schedule, utilizing the additional time only if needed. To the extent that it is possible, buffer time should be used for unanticipated events.

Close the Project

It is as important to close the project properly as it is to begin it properly. In traditional, long projects, the team never gets a sense of completion and accomplishment. The Close phase gives the team an opportunity to declare success, wrap up any loose ends, conduct a postmortem on the entire project, and even have a party.

In some organizations, projects seem to go on for years. When major changes occur in projects, it is important to close the project and initiate another one. Closing projects doesn't take much time, but the benefits can be substantial.

Summary

➤ The project management model is one of the foundations for a successful adaptive project.

➤ Preparation through good project initiation is key to both mountaineering adventures and software projects. Extra time spent in the beginning pays large dividends later.

➤ Project initiation involves choosing and preparing the project team, creating and assembling project mission data, and defining project structure items.

➤ The project approach consists of roles and responsibility, communication and collaboration, development practice, change management, learning, and progress assessment decisions.

➤ A very effective way to increase a project's speed is to start early.

➤ Planning the project starts with understanding the scope and requirements, and ends with creating a well-considered, scheduled, resource-loaded, risk-analyzed project plan.

➤ Project planning activities are

- Define the work.

- Develop the project schedule.

- Analyze the resource requirements.

- Assess project risk.

➤ Time-boxing is a structure for focusing on the most important mission characteristics and forcing engineering trade-offs.

➤ Time-boxing is not an excuse for low quality. It forces the project team and management to articulate the important dimensions of quality and their relative priorities.

➤ Developing a project schedule involves understanding the project's size, the cycle plan, and the resource availability.

➤ Risk analysis is an important part of cycle planning and ongoing project management. Risk analysis involves asking, "What do we not understand about this project?"

➤ Managing the project requires persistence, vigilance, and an ability to finish strong.

➤ The end, when staff is tired and tempers are short, is the time to remember the important practices. It is not the time to abandon reason and judgment.

➤ Change containment practices are more flexible than change control practices.

➤ It is important to close projects. Team members need a sense of closure.

CHAPTER 12
Dawdling, McLuhan, and Thin Air

This book is fundamentally about high speed, uncertainty, and high change. In the race to deliver a product quickly, however, it is not always necessary to go fast. There is a story about African bushmen guiding a European military excursion across a wide expanse of the African plain. In a race against time, the Europeans maintained a furious pace until the bushmen, almost in unison, paused, sat down, and refused to move. To the Europeans' pleading to resume the trek, the bushmen replied, "We have to wait for our souls to catch up."

Some things are not a function of speed, but of time and timing. What would your reaction be to an artist who said, "Between 3 and 4 P.M., I will paint the background trees; from 4 to 5, the flowers; from 5 to 8, the old barn in the center"? What would your response be to parents who allocated 6:30 to 7:00 P.M. each Wednesday and Friday to building relationships with their children? I imagine you would respond as I would, that creativity cannot be predicted and scheduled, that relationships cannot be built through regimented injections of attention. We respond to relationship events when the need arises, not when it is most convenient.

Just as in life, management of time and relationships is at the heart of successful Adaptive Software Development. To enable them to better manage relationships, many software developers have been introduced to Myers-Briggs personality typing, which facilitates an understanding of team dynamics. For example, Gerald Weinberg uses a form of Myers-Briggs personality-type indicators in his books on quality software management. Author Susan Campbell provides yet another classification of personality types in *Beyond the Power Struggle*. She describes Lovers and Warriors, Lookers and Leapers, Thinkers and Feelers, Spenders and Savers, and Scurriers and Dawdlers. The last of these, the dawdlers, exhibit behavior that can be important to software developers on extreme projects.

Dawdling

Dawdling, Campbell points out, is therapeutic. In the rush of high-speed projects, there are times, more frequently needed than we realize, when people ought to dawdle, to forget about doing, and just "be." There are times, especially when team members are in the groan zone, that anxiety is high and the end seems unattainably far down the road, when people just need to be together, to throw away the agenda, and to listen to each other.

"The world—the Western industrialized world, at least—seems to be speeding up, crowding out the dawdlers, crowding out the people who live life in the slow lane, the people who take time to listen and to notice and to ponder, the people who 'waste time in idle lingering' and find that such wasted time is well-spent. I fear that if the art of dawdling is lost in our world, we will have lost something precious," observed the Reverend Suzanne Spencer, then minister of Salt Lake City's South Valley Unitarian-Universalist Society, in a sermon given several years ago.

Clearly, time spent dawdling can give rise to the creative *aha!*, to the emergence of a novel solution, to the single, seemingly mundane event that causes a team to jell. The benefits to be derived from dawdling can't be predicted, planned, coerced, forced, or anticipated. Such successes often come from left field, generating emergent results from waves of intense data-gathering and analysis, interspersed with periods of quiet reflection. Collaboration and self-organization cannot always be hurried. Occasionally, we must slow down in order to progress more rapidly.

McLuhan

A couple of years ago, my colleague Steve Smith suggested that I test the Adaptive Software Development approach by applying Marshall McLuhan's four questions about new technology (McLuhan88, pp. 98–99). I thought for weeks about how best to apply McLuhan's questions to Adaptive Software Development: What does it enhance? What does it make obsolete? What does it bring back? What does it flip into? Finally, at a leisurely breakfast one Sunday morning, the answers began to emerge.

What Does Adaptive Software Development Enhance?

In the Preface, I identified Adaptive Software Development as an approach designed to

- offer emergent order as an alternative to the more prevalent practices of imposed order in building complex systems

- provide a series of models to help practitioners move from the conceptual to the actionable

- establish collaboration as the organizational vehicle for emergence

- provide a path for scaling adaptive development up to larger projects

- define Leadership–Collaboration, an adaptive management culture that describes the environment in which complex projects can be completed

It is my belief that every author wants to produce something new, something unique, something that enhances knowledge. But since the bulk of all knowledge is built on the foundations laid down by predecessors, a paradox arises. What is new? What is borrowed and integrated from others? The answer, for me, came from my remembering one of the fundamental characteristics in John Holland's definition of adaptive systems: They are developed from building blocks. Biological agents evolve by recombining existing building blocks until a different organism emerges. It is this recombination of building blocks that enables complex organisms to evolve into new forms, which are better able to prosper in their environment. In science, new building blocks are rare, but the rearrangement of existing building blocks provides an unlimited space for innovation.

Adaptive Software Development reorganizes existing building blocks and presents them from a new perspective. Complex adaptive systems theory is not new, but its application to organizational management has come about relatively recently. The concepts of iterative life cycles, customer focus groups, and mission statements also are not new, but I believe linking them to organizational learning theory and CAS presents them in a new way. Focusing on results and work*state* management is not new, but it goes against the tide of SEI-like process-improvement initiatives.

Self-organization, emergence, and collaboration are the central themes of Adaptive Software Development. Collaboration puts an emphasis on community, on trust and joint responsibility, on partnership with the customer rather than on an adversarial relationship. In climbing, when one's muscles are straining with lactic-acid buildup and one's thoughts are fuzzy from the altitude, the body reverts to older, deeply ingrained habits. With the ever-increasing speed of the Internet and mounting pressures on information technology at the century's end, there seems to be a regression to the *familiar*—to Command–Control management. More open, collaborative-style management is severely tested in the face of these reactionary pressures. When used properly, the Adaptive Software Development approach enhances and encourages the movement toward learning organizations, collaborative groups, and balancing the advantages of both competition and cooperation.

What Does Adaptive Software Development Make Obsolete?

I hope that the Adaptive Software Development approach makes obsolete the notion that linearity, predictability, arduous process improvement, and Command–Control are the only way to develop software. I hope it puts to rest forever the concept that a software *factory* in which humans assemble products as the speeding conveyer belt hypnotizes them into robotic behavior can ever be an effective way to produce software. As purveyors of modern methods in the Information Age, how did we ever let the word "factory" back into our conversation?

Let me be very clear—while people operating as robots is out, rigor is still part of the equation. It is ironic that ASD may even help chaotic organizations implement *appropriate* rigor. One of the reasons flexible, *ad hoc* organizations resist any rigor is that they perceive it as a one-way road to bureaucracy and loss of self. This belief results from their Accidental-versus-Monumental view of the world. They need another perspective on rigor.

In an optimizing environment, the objective of rigor is to establish more and more order. In such an environment, the view is that if some rigor is good, more rigor is better. In an adaptive environment, however, the objective of rigor is very different. There, the goal must be to perch the organization on the edge; to provide enough balance to enable even greater creativity, innovation, and emergence; and to contribute a counterweight when true chaos looms too ominously. An adaptive culture understands that rigor is a balancing mechanism, not an end in itself.

What Does Adaptive Software Development Bring Back?

At a philosophical level, an adaptive approach brings back a sense of self, a sense of purpose. It seems to me that optimization practices such as the Capability Maturity Model and Business Process Reengineering reduce an individual's sense of contribution and worth, even though this is not the stated intent. Individuals are just part of the process, replaceable at will.

Adaptation, however, stresses the importance of individuals and their connections. Policies, procedures, and rules are minimized, since their intent here is to offer boundaries within which individuals can operate creatively. Optimization stresses processes—prescriptions for actions—which seek to limit behavior within a narrow range of acceptability. Adaptation stresses patterns and interrelationships and expanding behavior—setting the limits for our creativity at the outer boundaries, not at the inner ones.

Optimization proponents may consider the above disingenuous, arguing that their approach has a "people component." However, I would contend that their frameworks and principles put people second.

What Does Adaptive Software Development Flip Into?

Marshall McLuhan indicated that every new technology "flips" into something, or some things, that the originators never intended. My fear is that adaptive development enthusiasts may use the ASD approach as an excuse to "flip" into anarchy. Some may interpret adaptive concepts as a justification for an "adhocracy"—an environment governed by *cowboy, ad hoc* coding. A careful reading of this book *should* disabuse people of this idea, but then, a reader always is free to pick out what he or she will, as I do when reading someone else's work.

Nevertheless, I contend that process improvement and Business Process Reengineering have gone too far—they have flipped into organizational systemization, a state antithetical to innovation and emergence. Taken too far, adaptation flips into the opposite realm—one that breeds anarchy, fuzzy thinking, a lack of direction, and, ultimately, a lack of results.

As to other flippings—well, I'll just stand back and wait to see what emerges.

Organizational Growth

Some readers may question whether complex adaptive systems theory is anything more than an interesting metaphor for a different style of organizational behavior. The same could be asked about the traditional metaphors taken from the science of Newton and Darwin. Whether or not the old science is applicable to organizations, we have used it for more than one hundred years. Whether they are good metaphors or not, the concepts of survival of the fittest, optimization, and determinism have been pervasive in management. The old science, and the old management, have stood the test of time.

But we are in a new time, and we are in need of a new foundation, which I believe is one of *adaptation* and *arrival of the fittest*. Traditional approaches do not help us survive and prosper in high-change, high-speed environments. The evidence is growing that newer, different styles of management are working more effectively. The more we learn about underlying principles and why the new styles work, the better we can design organizations to respond to turbulent environments in the future.

Complex adaptive systems theory is relatively new. Although the seeds have been germinating for several decades, only within the last ten years have they developed. It may be too early to forecast CAS's impact on management, but it is not too early to speculate. And the speculation, from writers like Margaret Wheatley and Ralph Stacey and from organizations like the Ernst & Young Center for Business

Innovation and the Santa Fe Center for Emergent Strategies, suggests that the new science has enormous potential.

The application of this new science to management, as put to use in this book, can be summarized by four statements. The first two of these summary statements deal with people and environments:

> **People want to succeed; given the right environment, people motivate themselves to heroic achievements.**

> **The right environment for extreme projects is one of adaptation through self-organization.**

The idea that people in the right environment achieve success is central to this book. To questions like, "Why do people climb mountains?" or "Why do people engage themselves in software projects?" or "Why do some people, and some teams, push themselves to heroic limits?" the casual reply is, "Because they are there," but there are deeper reasons.

If some company coerced employees to climb mountains (as companies at times coerce people into working on death-march projects), it would be accused of abuse, exploitation, and worse. Long days, physical exhaustion, life-threatening danger, interlaced with periods of adrenaline rush—the only comparable job might be that of an infantry soldier in war. OSHA would not approve! People can't be commanded to climb mountains, but thousands and thousands do it of their own free will—in fact, they often pay a great deal of money for the privilege.

"You may be able to kick people to make them active, but not to make them creative, inventive, and thoughtful."
—T. DeMarco and T. Lister [1999], p. 9.

Nor can people be commanded to carry out the heroic measures necessary in extreme software projects. Commenting in a Letter to the Editor of *American Programmer* regarding an issue on death-march projects (DeMarco97), Tom DeMarco expressed the view that forcing people through fear or intimidation to participate on a death-march project is the worst form of ethical lapse. DeMarco compares death-march projects to the actual Death Marches of Burma during World War II. Extreme projects should not be death marches.

People sign on to extreme projects for the same reasons they climb mountains—because of the challenge, the adrenaline rush, the sense of purpose, to accomplish something few others care to confront, to master something beyond the norm. People can be assigned to such projects, but they perform heroic feats as the result of an inner fire.

Project teams that burn with inner fire are not commanded, they are led. They are not controlled, they collaborate. They are not deterministic, linear, and stable; they are chaotic, nonlinear, anxiety-ridden,

"During my thirty-four-year tenure as a climber, I'd found that the most rewarding aspects of mountaineering derive from the sport's emphasis on self-reliance, on making critical decisions and dealing with the consequences, on personal responsibility."
—J. Krakauer [1997], pp. 167–68.

and exuberant. Such teams make many, many mistakes—but their successes far outweigh the mistakes, which serve as their motivation for learning how to succeed. Adaptive environments stand a much better chance of creating teams burning with inner fire.

Adaptation does not mean chaos. From complex adaptive systems, we know that creative, innovative results emerge—not from chaos, but from the transition zone at the edge of chaos. The idea that leading teams to the edge, keeping them there for extended periods, and scaling smaller successes up to larger efforts is an important objective expressed in this second pair of observations:

An iterative cycle of speculating, collaborating, and learning establishes a framework for adaptation.

Self-organization on a large scale depends on one's ability to manage information flows.

Adaptation takes more energy than many expect. Being willing to change and being willing to learn are not the same. True learning is always an emotional experience, not an analytical one—and therefore it is difficult for many software engineers, whose careers have been built on an analytical view of the world. Embracing an adaptive philosophy means first opening oneself up to admitting that we live in an indeterminate world, one in which chance and interactions with other players shape our future.

Embracing an adaptive philosophy means rethinking a century or more of management practices built on the foundations of the old science. Command–Control management and fierce competition are so ingrained in our culture, in part because of its *scientific* derivation. Only an equally strong *new science* can support the cultural changes upon which adaptation depends.

In extreme environments, the key to success is our ability to learn rapidly about what we do not understand. This learning requires that we have an open, inquiring perspective and the strength to embrace mistakes; it assumes iterative learning cycles to test for understanding. Understanding depends upon diversity—it means testing and learning from customers, suppliers, competitors, and other departments and work groups within one's own organization. We must release partially completed work products to these diverse stakeholders, and then we will learn—from both successes and failures. To get it *right* the last time, we must learn from interim releases by doing it *wrong* the first time.

Iteration and an adaptive philosophy are keys to understanding and learning. Iteration acknowledges that every deliverable component is subject to change.

Speed is a function of many variables—skill, focused energy, dedicated resources—but one essential technique for increasing development speed is *concurrency*. Without concurrent feature-set development and without overlapping requirements, design, and coding, products simply cannot be developed at competitive speeds.

Multiple iterations are key to learning, but they cause change. Concurrency is key to speed, but it causes change also. Iteration on top of concurrency results in chaos or products, depending on how it is managed. Command–Control management will not work in this environment; the information exchange rate is just too high. Without collaboration and decision-making at the work-group level, projects tip into chaos.

As software developers, we live in a world of abstraction and information. Complex adaptive systems theorists focus on information, as defined by three of Ralph Stacey's control parameters: rate of information flow, degree of diversity, and richness of connectivity. These same parameters help us design the collaborative structure upon which Adaptive Software Development depends. "It all comes down to information," as George Johnson's quote in the sidebar contends.

The task of organizing a group of individuals into a high-performance feature team is difficult. The task of organizing a group of individual feature teams into a high-performance product team is daunting. Self-organization succeeds when the right information flows to the right people at the right time—too little information and the organism starves; too much causes a kind of obesity, slowing response time and dulling the fire.

The collaboration structure is a mechanism to facilitate self-organization within and across teams—particularly in those spread over time and space. The job of this structure is to create the "high context" of small, collocated groups over large, distributed, virtual teams and to manage the flow of high-content information.

At a workshop sponsored by the Santa Fe Center for Emergent Strategies, one of the Center's managing directors Mike Simmons talked with me about the applicability of complexity theory to high-technology businesses. He commented that the latter already seem to employ many adaptive practices. Our conversation started me thinking about the applicability of adaptive concepts to both Monumental

"We think of matter and energy as fundamentals—we can feel the heft of a rock or the jolt of electricity. Information seems subjective. Yet why should what we know through our bodies be more fundamental than what we know through our brains? In the end, we only know about matter and energy through the signals sent by our senses—our eyes, ears, noses, the receptors in our skin. It all comes down to information."
—G. Johnson [1995], p. 130.

and Accidental organizations and about these organizations' contrasting characteristics. Stated simply,

> **Monumental organizations know how they want to be, and they don't want to change.**

> **Accidental organizations know how they don't want to be, but not how they want to be.**

Monumental organizations use a fundamentally deterministic model. They believe in prediction, linearity, comfort, and Command–Control. As the world spins faster, these organizations need to change, but it seems the faster the spinning, the more tightly they hold on to keep from being thrown off. This only assures they will be jettisoned at maximum velocity.

Accidental organizations, as they grow from small, collocated, innovative work groups, *know how they do not want to be*—they don't want to be bureaucratic, stuffy, or overly constrained. But the model they have for growth is based on an overwhelming body of management knowledge rooted in the deterministic, linear, mechanistic view of organizations as machines. I've worked with software companies that have had a very difficult time putting mid-level management training together. Much of what is *available* seems irrelevant to how they actually operate. Thus, management training is reduced to a set of new "rules" that seem to work, but that at the same time do not have much overall context and depth—they are *rules* but not *principles*.

Many Accidental organizations, even larger ones, *do not know how they want to be*. They do not have a model as widely known and articulated as the optimization model. They rebel against too many rules, but the issue is not really about rules. It is about whether organizations are machines or living systems. Starting with the premise that organizations are living organisms and, furthermore, that they are *complex adaptive systems* provides a new foundation for *how they want to be*.

Both Monumental and Accidental organizations succeed in delivering products, while causing grave sickness to the organization—they just do it in different ways. Getting the product out at any cost (to people) is just as prevalent, if not more so, in high-tech companies as it is in lower-tech ones. Monumental organizations stifle innovative people; Accidental organizations exploit them. The organisms die in different ways, but they still die.

Surviving in Thin Air

Shipping products and creating healthy organisms are not antithetical objectives. Achieving a goal, whether a mountain summit or a shipped product, can have a tremendous, positive, euphoric effect on a person. Training for and running a marathon are extremely difficult activities, yet the impact on most people who achieve this goal is overwhelmingly positive. In the same way, working six months of sixty-hour weeks (training) and six weeks of eighty-hour weeks (running the race) to bring a product to a customer (finishing the race) can substantially increase the health of a project team, or it can make it very sick.

Self-reference is a living organism's inner mission—its sense of purpose that originates from within rather than being imposed from without. For many species, self-reference is directed to survival, but in humans it is more. Self-reference, whether for an individual, a project team, or a company, goes beyond physical survival and procreation to an inner need to pursue a *purpose* for surviving. Whether it is Jefferson's and Lewis's grand views of a continental United States, a Peace Corp volunteer's will to help others, or an Internet software team's desire to use the power of communications and collaboration to assault bureaucracy, self-reference is a powerful force.

Motivation by fear, intimidation, power, or bribery is antithetical to self-reference. Motivation by appeal to physical survival (that is, at the very least, what someone must do to keep a job) *does* work. But, as the bushmen in the opening story of the chapter might have observed, it doesn't provide for the soul.

Two projects, identical by every external measure—long hours, hard work, frustration, anxiety, euphoria, depression—can be dramatically different. One can feed the team's quest for self-reference, producing a healthy, vital, vibrant (although tired) group. The second can produce a burned-out, empty, morose group of people who just want to pick up a paycheck and go home. Same physical characteristics, different outcome.

"More than any other science principle I've encountered, self-reference strikes me as the most important."
—M. Wheatley [1992], p. 146.

The lure, for me at least, of applying complex adaptive systems concepts to software projects is that doing so provides a conceptual foundation supporting both *survival*, which produces concrete results, and *purpose*, which feeds the spirit.

It would be difficult to employ adaptive development without understanding the difference between survival and purpose. In *People-*

ware: Productive Projects and Teams (DeMarco99), Tom DeMarco and Tim Lister discuss the signs of a jelled team:

> *Things that matter enormously prior to jell (money, status, position for advancement) matter less or not at all after jell. . . . Jelled teams are usually marked by a strong sense of identity. . . . a sense of eliteness . . . joint ownership of the product. . . . obvious enjoyment.* —T. DeMarco and T. Lister [1999], p. 127.

A jelled team's sense of "identity," of purpose beyond delivering a product, produces a juggernaut. No amount of external motivation causes a team to jell. Jelling comes from within. Therefore, management practices built around the living systems metaphor have a better chance of tapping into the power of self-reference.

Jon Krakauer's *Into Thin Air* (Krakauer97) tells a vivid, first-person account of the climbing expedition on Mount Everest in 1996 in which five people lost their lives. (Seven climbers from other expeditions also died in their attempts.) Climbing Everest is an extreme project, similar to the software projects that are the focus of this book. Software projects may not kill people, but they can cause burnout and severe mental anguish. In extreme projects, the air is thin, the margin of error small, the weather conditions harsh, and the mountain steep and hazardous. Extreme projects are not a place to follow others, but a place for deep personal responsibility, a place where judgment is critical because the consequences can be devastating.

Ultimately, succeeding on extreme projects is about personal responsibility. Author James Bach (Bach95) writes about heroics—specifically, all the heroic acts it takes to produce a product in extreme conditions. Bach is not referring to the effort of a single Herculean figure, but to the hero in all of us that drives our dedication and gives us the will to persist in the face of great odds. Chapter 1 of this book recommended measuring success on extreme projects in terms of three criteria:

- The product gets shipped.

- The product approaches its mission profile, a prioritized combination of scope, schedule, resource, and defect levels.

- The project team is healthy at the end.

By focusing on these criteria while providing concepts and practices that address both survival and purpose, the Adaptive Software Development approach assuredly will help project teams succeed.

Bibliography

Adams, Scott. *The Dilbert Principle.* New York: HarperBusiness, 1996.

Ambrose, Steven E. *Undaunted Courage: Meriwether Lewis, Thomas Jefferson, and the Opening of the American West.* New York: Simon & Schuster, 1996.

> For history buffs, an excellent and thoroughly enjoyable biography of Meriwether Lewis, including of course, the famous expedition.

Andreas, Steve, and Charles Faulkner. *NLP: The New Technology of Achievement.* New York: Quill, 1994.

> A good introduction to the field of Neuro-Linguistic Programming.

Arthur, W. Brian. "Increasing Returns and the New World of Business." *Harvard Business Review,* Vol. 74, No. 4 (July-August 1996), pp. 100–110.

> Arthur is one of the founders of the Santa Fe Institute, and a leader in the application of complexity theory to the realm of economics. This article provides a great start for exploring economics from a different perspective.

Austin, Robert D. *Measuring and Managing Performance in Organizations.* New York: Dorset House Publishing, 1996.

Bach, James. "The Challenge of Good-Enough Software." *American Programmer*, Vol. 8, No. 10 (October 1995), pp. 3–11.

> Bach is a spokesperson for the "anti-SEI, let's be reasonable" school of development. The response by the process-rigor community shows that he hits a sore nerve.

Bayer, Sam, and Jim Highsmith. "RADical Software Development." *American Programmer*, Vol. 7, No. 6 (June 1994), pp. 35–42.

Bennis, Warren. *On Becoming a Leader*. Reading, Mass.: Addison-Wesley Publishing Co., 1989.

_____, and Patricia Biederman. *Organizing Genius: The Secrets of Creative Collaboration*. Reading, Mass.: Addison-Wesley Publishing Co., 1997.

> An informative and readable book from two of the best-known writers on leadership. This book explores leadership through examining seven, high-profile, *genius*-laden groups, including the Disney animation studio, Lockheed's famous Skunk Works for airplane design, and Xerox's Palo Alto Research Center (PARC).

Boehm, Barry. "A Spiral Model of Software Development and Enhancement." *IEEE Computer*, Vol. 66, No. 5 (May 1988), pp. 61–72.

_____. "Software Risk Management: Principles and Practices." *IEEE Software*, Vol. 69, No. 1 (January 1991), pp. 32–41.

_____, Alexander Egyed, Julie Kwan, Dan Port, Archita Shah, and Ray Madachy. "Using the WinWin Spiral Model: A Case Study." *IEEE Computer*, Vol. 31, No. 7 (July 1998), pp. 33–44.

Brown, Shona L., and Kathleen M. Eisenhardt. *Competing on the Edge: Strategy as Structured Chaos*. Boston: Harvard Business School Press, 1998.

> For practical, usable ideas about the application of complexity concepts to business strategy, this book is high on my list. In addition to strategies for balancing on the edge of chaos, the authors also discuss the strategies of balancing on the edge of time and time pacing. Very good ideas for high-speed, turbulent market environments.

Charette, Robert N. *Software Engineering Risk Analysis and Management*. New York: McGraw-Hill, 1989.

_____. "A Software Management Special Report." *Software Management* (October 1993), pp. 5–11.

Christensen, Clayton. *The Innovator's Dilemma: When New Technologies Cause Great Firms to Fail.* Boston: Harvard Business School Press, 1997.

A must-read for anyone working in fast-moving, high-change, high-technology environments who is interested in innovation and how to sustain innovative organizations over time. Some technologies "disrupt" the status quo in ways that only time can sort out. I believe some management practices "disrupt" the status quo also and that practices that have evolved from complex adaptive systems concepts are an example of disruptive management practices.

Coleman, David. *Groupware: Collaborative Strategies for Corporate LANs and Intranets.* Englewood Cliffs, N.J.: Prentice-Hall, 1997.

A compendium of the state of the art in collaborative groupware as of early 1997. Focused more on technical aspects, the book contains a good framework for the different kinds of groupware.

Conner, Daryl R. *Managing at the Speed of Change.* New York: Villard Books, 1992.

An interesting book on change management from one of the best-known consultants in the field.

Constantine, Larry. *Constantine on Peopleware.* Englewood Cliffs, N.J.: Prentice-Hall, 1995.

A collection of Constantine's essays, primarily from his column in *Software Development.* Lots of good ideas for team leaders and project managers. Constantine has a jaunty style, which makes for fun reading.

Coveney, Peter, and Roger Highfield. *Frontiers of Complexity: The Search for Order in a Chaotic World.* New York: Random House / Fawcett Columbine, 1995.

Crichton, Michael. *The Lost World.* New York: Alfred A. Knopf, 1995.

Cusumano, Michael A., and Richard Selby. *Microsoft Secrets.* New York: The Free Press, 1995.

The best of the books so far on the Microsoft Process. Good information, readable, but overly detailed (for my taste, at least) in some chapters.

Davenport, Thomas, and Laurence Prusak. *Working Knowledge: How Organizations Manage What They Know.* Boston: Harvard Business School Press, 1998.

De Geus, Arie. *The Living Company: Habits for Survival in a Turbulent Business Economy.* Boston: Harvard Business School Press, 1997.

> A very readable book that treats companies like living organisms. De Geus explores what encourages and discourages companies from living longer lives.

DeMarco, Tom. *Why Does Software Cost So Much?* New York: Dorset House Publishing, 1995.

> This book contains a series of essays as only DeMarco can write them. Anyone interested in measurement should read Chapter 2, "Mad About Measurement." DeMarco asserts that most measurement programs are instituted not to discover facts or steer our actions, but to modify human behavior, and he deplores this trend. On a personal note, his description of the hours and hours and hours he spends editing his own writing helped me get through this book.

_____. "Letter to the Editor." *American Programmer,* Vol. 10, No. 4 (April 1997), p. 2.

_____, and Timothy Lister. *Peopleware: Productive Projects and Teams,* 2nd ed. New York: Dorset House Publishing, 1999.

> Buy it! Read it! An all-time best-seller, first published in 1987.

Doyle, Michael, and David Straus. *How to Make Meetings Work.* New York: Jove Books, 1976.

> Around for twenty-plus years, this is still one of the better books on meeting techniques.

Frick, Don M., and Larry C. Spears, eds. *On Becoming a Servant Leader: The Private Writings of Robert K. Greenleaf.* San Francisco: Jossey-Bass Publishers, 1996.

Gause, Donald C., and Gerald M. Weinberg. *Exploring Requirements: Quality Before Design.* New York: Dorset House Publishing, 1989.

Gell-Mann, Murray. *The Quark and the Jaguar: Adventures in the Simple and the Complex.* New York: W.H. Freeman and Company, 1994.

> Quantum physicist Gell-Mann is a winner of the Nobel Prize in Physics for his discovery of quarks. This book is his attempt at explaining complexity to the uninitiated. Gell-Mann's book for nonscientists is still heavy going. However, as he is one of the founding members of the Santa Fe Institute and bearer of the complex adaptive systems exploration flame, it is worthwhile to read what he has to say.

Gilb, Tom. *Principles of Software Engineering Management.* Wokingham, England: Addison-Wesley Publishing Co., 1988.

> Gilb pioneered much of the work on evolutionary software development, as well as wrote on other areas of software management. His ideas are still relevant today.

Glass, Robert L. *Software Creativity.* Englewood Cliffs, N.J.: Prentice-Hall, 1995.

> A book about the Yin and Yang of development. Glass explores the relationship between rigorous development practices and creativity.

Goddard, Dale, and Udo Neumann. *Performance Rock Climbing.* Mechanicsburg, Penn.: Stackpole Books, 1993.

> If you are a rock climber, and want to improve your skills, mental conditioning, strength, and all-around climbing technique, this eminently readable text is one of the best sources.

Goldratt, Eliyahu M. *The Goal.* Great Barrington, Mass.: The North River Press, 1984.

_____. *Critical Chain.* Great Barrington, Mass.: The North River Press, 1997.

Goldstein, Jeffrey. *The Unshackled Organization: Facing the Challenge of Unpredictability Through Spontaneous Reorganization.* Portland, Ore.: Productivity Press, 1994.

Gould, Stephen Jay. *Wonderful Life: The Burgess Shale and the Nature of History.* New York: W.W. Norton and Company, 1989.

> Gould's best. The Burgess Shale discovery was overlooked for years, and then became one of the most prolific sources of undis-

covered species and entire biological phyla. If you have any interest in biological evolution, this is a *wonderful* book. Gould's underlying thesis (opposite to Stuart Kauffman's) is that evolution is accidental and unpredictable, so that if the tape of life were rewound and replayed, the results (you and me, for example) would be completely different. That is, you would not be reading this sentence I did not write.

Haeckel, Stephan H. *Adaptive Enterprise: Creating and Leading Sense-and-Respond Organizations.* Boston: Harvard Business School Press, 1999.

Stephan Haeckel, director of strategic studies at IBM's Advanced Business Institute, provides must-reading for large organizations that are trying to become more adaptive. Growing out of the recent sense-and-respond strategic framework (in which organizations "sense" what customers want and then "respond" appropriately, in contrast to a traditional make-and-sell strategy in which companies "make" products and then attempt to sell them to customers), *Adaptive Enterprise* fills in a critical piece in using complexity concepts in organizational strategy.

Highsmith, Jim. "Software Ascents." *American Programmer,* Vol. 5, No. 6 (June 1992), pp. 20–26.

Adapted portions of "Software Ascents" appear in Chapter 1 of this book.

_____. "Messy, Exciting, and Anxiety-Ridden: Adaptive Software Development." *American Programmer,* Vol. 10, No. 4 (April 1997), pp. 23–29.

_____. "Order for Free." *Software Development,* Vol. 6, No. 3 (March 1998), pp. 78–80.

Adapted portions of "Order for Free" appear in Chapters 7 and 10 of this book.

_____. "Thriving in Turbulent Times." *Application Development Strategies,* Vol. 10, No. 12 (December 1998), pp. 1–16.

_____, and Lynne Nix. "Mission Possible: Feasibility Analysis." *Software Development,* Vol. 4, No. 7 (July 1996), pp. 40–45.

Hof, Robert D. "Netspeed at Netscape." *Business Week* (February 10, 1997), pp. 48–51.

Holland, John H. *Hidden Order: How Adaptation Builds Complexity.* Reading, Mass.: Addison-Wesley Publishing Co., 1995.

> A short, dense book. Holland defines the common attributes of any complex adaptive system, be it for an ant hive, a rain forest ecosystem, or a human organization. His work includes the development of simulation models to investigate the properties exhibited by all complex adaptive systems. Not easy, but another must-read for those interested in a serious study of complexity.

_____. *Emergence: From Chaos to Order.* Reading, Mass.: Addison-Wesley Publishing Co., 1998.

> For a book explicitly not about management and organizations, Holland's latest work is destined to have significant impact in that arena. Holland explores emergence from the perspective of a scientist, and limits his conclusions to systems in which there are descriptive rules or laws. While not ruling out the applicability to other areas (like organizations), Holland leaves extension of the concepts to others or to the future. But one cannot read this book without acquiring a deep understanding of emergence and its potential for management. More readable to the nonscientist than his prior book, *Emergence* is destined to become a seminal work. This, and Stuart Kauffman's *At Home in the Universe,* should be at the top of the list for anyone wanting a background in the science of complexity.

Humphrey, Watts S. *Managing the Software Process.* Reading, Mass.: Addison-Wesley Publishing Co., 1989.

> For those who want a better understanding of the deterministic, optimization approach to software development, Humphrey is the best-known spokesperson.

Johnson, George. *In the Palaces of Memory: How We Build the Worlds Inside Our Heads.* New York: Alfred A. Knopf, Inc., 1991.

> One of the most complex biological phenomena is the human mind. Johnson creates a historical voyage into explaining how memory works, and visits neurobiology, computer science, philosophy, psychology, and other areas to examine memory from a variety of perspectives. Although it is not a book specifically about complexity theory, the reader is left with a rich understanding of a realm that complexity theory helps us understand.

_____. *Fire in the Mind: Science, Faith, and the Search for Order.* New York: Vintage Books, 1995.

> A marvelous, fascinating book! Johnson is a science writer for *The New York Times.* This book is an exploration of the science of complexity, but he weaves his exploration of quarks, complex adaptive systems, information physics, and evolutionary biology with tales of the cosmologies of the Tewa Indians living around Santa Fe, New Mexico. The core questions are about reality, myth, science, and faith. Johnson's exploration of the questions is masterful.

Johnson, Jim. "Creating Chaos." *American Programmer,* Vol. 8, No. 7 (July 1995), pp. 3–7.

Jones, Capers. *Applied Software Measurement.* New York: McGraw-Hill, 1992.

> The classic in software measurement. A must for every software engineer's bookshelf. A second edition is now available.

_____. *Patterns of Software Systems Failure and Success.* Boston: International Thomson Computer Press, 1996.

Kaner, Sam. *Facilitator's Guide to Participatory Decision-Making.* Philadelphia: New Society Publishers, 1996.

> On my *must-read* list. In understanding group dynamics and learning about facilitation skills, Kaner is at the top of the heap. This book is packed with practical techniques, and organized into a very friendly and usable conceptual framework. Forget the facilitation books that dwell on how many pencils to bring for each participant—get this book instead. Kaner's *groan zone* is a great phrase for a perpetually present organizational dynamic.

Katzenbach, Jon R., and Douglas K. Smith. *The Wisdom of Teams: Creating the High-Performance Organization.* Boston: Harvard Business School Press, 1993.

> The hot management book of 1993 and 1994. Provides a useful framework for anyone interested in team issues.

Kauffman, Stuart. *The Origins of Order: Self-Organization and Selection in Evolution.* New York: Oxford University Press, 1993.

> This book is more difficult than *At Home in the Universe.* Seven hundred pages on wide-ranging subjects, from evolution to cellular automata to ontogeny. I must admit I haven't made it all the way through yet.

_____. *At Home in the Universe: The Search for the Laws of Self-Organization and Complexity*. New York: Oxford University Press, 1995.

> Along with Gell-Mann, Holland, Arthur, and a few others, Kauffman helped shape the emergence of the field of complex adaptive systems. His books are often difficult going, especially for readers without a solid foundation in biology. Some knowledge of Kauffman's work in Genetic Network simulations and his conclusions about the nature of order in the world is very helpful in understanding the nature of emergence, self-organization, natural selection, and the structure of systems at the edge of chaos. Many of the other books on complex adaptive systems are syntheses, or explanations, of the ideas of Kauffman and the others mentioned above. My recommended sequence—Kauffman, Holland, Arthur, Gell-Mann.

Kelly, Kevin. *Out of Control: The New Biology of Machines, Social Systems, and the Economic World*. Reading, Mass.: Addison-Wesley Publishing Co., 1994.

> Ask yourself, "What would I expect from the editor of *WIRED* magazine?" and you have some idea of the range of material in *Out of Control*. Kelly is WIRED and the current is flowing. Ants, robots, networks, Post-Darwinism, ecology, the Biosphere project, and much more confront the reader and build a growing gestalt of what the future may hold for social and economic ecosystems. If you enjoy writers like Gleick *(Chaos)* and Hofstadter *(Gödel, Escher, Bach: An Eternal Golden Braid)*, you should enjoy this book.

_____. *New Rules for the New Economy*. New York: Viking Press, 1998.

> Kelly's more recent work and, while worthwhile, much less interesting to me.

Keuffel, Warren. "Just Doing It." *Software Development*, Vol. 5, No. 11 (November 1997), pp. 31–32.

> This piece appeared in Keuffel's "Tools of the Trade" column in *Software Development* and reports on Nike's use of adaptive techniques on a project.

_____. "People-Based Processes: a RADical Concept." *Software Development*, Vol. 3, No. 11 (November 1995), pp. 31–37.

Kidder, Tracy. *The Soul of a New Machine*. Boston: Little, Brown & Co., 1981.

Krakauer, Jon. *Into Thin Air: A Personal Account of the Mt. Everest Disaster*. New York: Villard Books, 1997.

> A best-seller and a fascinating account of the limits human beings can push themselves to, and the consequences of that pushing.

Kramer, Matt. *Making Sense of Wine*. New York: William Morrow & Co., 1989.

> For wine lovers and even for those who partake only occasionally, an excellent excursion into the sometimes overly esoteric realm of wine drinking. Includes an interesting take on complexity and how it affects our senses.

Leonard, Andrew. *BOTS: The Origin of New Species*. San Francisco: HardWired, 1997.

> Bots, Robots, Cancelbots, Bots gone bad, Chatterbots, flaky Bots, fun Bots, stupid Bots, scary Bots—Leonard reviews the history and potential future of Bots, digital robot species inhabiting cyberspace. His book makes one realize that Asimov was wrong—there is no first law of robot behavior. Bots and robots are built by humans, and the closer their characteristics are to a human's, the more human they become in both positive and negative ways.

Levy, Steven. *Artificial Life: A Report from the Frontier Where Computers Meet Biology*. New York: Vintage Books, 1992.

> Artificial Life, or A-Life, is a fascinating adjunct to the field of complex adaptive systems. Many of the A-Life simulations provide a laboratory for understanding complexity. Levy traces the history, which is somewhat ancient history by now, of the A-Life movement. Worthwhile.

Lewin, Roger. *Complexity: Life at the Edge of Chaos*. New York: Macmillan Publishing Company, 1992.

Lipnack, Jessica, and Jeffrey Stamps. *The TeamNet Factor*. Essex Junction, Vt.: Oliver Wright, 1993.

————. *Virtual Teams*. New York: John Wiley & Sons, 1997.

McCarthy, Jim. *Dynamics of Software Development*. Redmond, Wash.: Microsoft Press, 1995.

McConnell, Steve. *Rapid Development*. Redmond, Wash.: Microsoft Press, 1996.

A well-written compendium of solid, practical software-development practices oriented toward accelerating delivery.

McLuhan, Marshall, and Eric McLuhan. *Laws of Media: The New Science*. Toronto: University of Toronto Press, 1988.

Maguire, Steve. *Debugging the Development Process*. Redmond, Wash.: Microsoft Press, 1994.

Moore, Geoffrey A. *Crossing the Chasm*. New York: HarperBusiness, 1991.

_____. *Inside the Tornado*. New York: HarperBusiness, 1995.

If you are in the software business, and want to stay for a while, do not pass up Moore's books. They explain what everyone in high-technology markets needs to know about markets, and about how markets affect all people in the organization. Well-written, these books will help developers understand what marketing departments are about.

Orsburn, Jack, Linda Moran, Ed Musselwhite, and John Zenger. *Self-Directed Work Teams: The New American Challenge*. Homewood, Ill.: Business One Irwin, 1990.

Peters, Thomas J., and Robert H. Waterman, Jr. *In Search of Excellence: Lessons from America's Best-Run Companies*. New York: Harper & Row, 1982.

Peters, Tom. *Thriving on Chaos*. New York: Alfred A. Knopf, 1987.

Petzinger, Thomas. *The New Pioneers: The Men and Women Who Are Transforming the Workplace and Marketplace*. New York: Simon & Schuster, 1999.

Petzinger, who writes the Friday "Front Lines" column for *The Wall Street Journal*, chronicles the new pioneers of business and how they are transforming the economy. His stories run the gamut from the LDS Hospital in Salt Lake City to the multi-billion-dollar Koch Industries, and provide case after case of how understanding businesses as complex adaptive systems is changing our mental model of business. This book should be very high on your list.

Poundstone, William. *Prisoner's Dilemma.* New York: Anchor Books, 1992.

> A readable book about Game Theory in general, but also a brief biography of John von Neumann. Von Neumann, who computer scientists know for his pioneering work in programming, was a mathematical genius who did pioneering work in Game Theory and cellular automata as well as pure mathematics. He was part of the brain trust in the Manhattan Project during World War II.

Rebello, Kathy. "Inside Microsoft: The Untold Story of How the Internet Forced Bill Gates to Reverse Course." *Business Week* (July 15, 1996), p. 56.

Rummler, Geary A., and Alan P. Brache. *Improving Performance: How to Manage the White Space on the Organization Chart.* San Francisco: Jossey-Bass Publishers, 1990.

> One of the better management books on organizational performance improvement.

Schaffer, Robert H., and Harvey A. Thomson. "Successful Change Programs Begin with Results." *Harvard Business Review,* Vol. 70, No. 1 (January-February 1992), pp. 80–89.

> A breath of fresh air in the smog of process mavens.

Schrage, Michael. *No More Teams: Mastering the Dynamics of Creative Collaboration.* New York: Currency Doubleday, 1989.

> Schrage's definition of collaboration is my favorite, and the ideas presented in the early chapters of the book are well worth reading.

Senge, Peter. *The Fifth Discipline: The Art & Practice of the Learning Organization.* New York: Currency Doubleday, 1990.

> A classic, maybe *the* classic, on organizational learning. I must admit I avoided the book for several years because it was too *in.* Although the area is somewhat faddish, there is tremendous substance behind the fad. High on my recommended list.

_____, Art Kleiner, Charlotte Roberts, Richard B. Ross, and Bryan J. Smith. *The Fifth Discipline Fieldbook.* New York: Doubleday, 1994.

Shafritz, Jay M., and J. Steven Ott, eds. *Classics of Organizational Theory,* 2nd ed. Pacific Grove, Calif.: Brooks/Cole Publishing, 1987.

Sherman, Howard, and Ron Schultz. *Open Boundaries: Creating Business Innovation Through Complexity*. Reading, Mass.: Perseus Books, 1998.

> A book about ideas, emergence, complexity, and innovation. A bit on the philosophical side, but full of insights to jog the thinking process.

Smith, Bradley J., Nghia Nguyen, and Richard F. Vidale. "Death of a Software Manager: How to Avoid Career Suicide Through Dynamic Software Process Modeling." *American Programmer*, Vol. 6, No. 5 (May 1993), pp. 10–17.

Smith, Preston, and Donald Reinertsen. *Developing Products in Half the Time: New Rules, New Tools*, 2nd ed. New York: John Wiley & Sons, 1997.

> Although this is a book about products in the manufacturing field, not about software products, it presents solid, practical advice on how to speed up the development process for all kinds of products. Well worth the read.

Sorensen, Reed. "A Comparison of Software Development Methodologies." *Crosstalk* (January 1995).

Stacey, Ralph D. *Complexity and Creativity in Organizations*. San Francisco: Berrett-Koehler, 1996.

> Although somewhat academic and staid, Stacey's book provides one of the most cogent arguments for classifying business organizations as complex adaptive systems. Where Margaret Wheatley's treatment of the topic is more philosophical and prosaic, Stacey builds a better conceptual foundation.

Stegner, Wallace. *The American West as Living Space*. Ann Arbor, Mich.: University of Michigan Press, 1987.

> Pulitzer prizewinner Stegner, who excels in both fiction and non-fiction, spent a lifetime trying to understand and write about the West. Reading Stegner is a real treat.

Tapscott, Don. *The Digital Economy: Promise and Peril in the Age of Networked Intelligence*. New York: McGraw-Hill, 1996.

Taylor, Frederick W. "The Principles of Scientific Management." In Shafritz and Ott, pp. 66–81 (originally published in *Bulletin of the Taylor Society*, December 1916).

Treacy, Michael, and Fred Wiersema. *The Discipline of Market Leaders.* Reading, Mass.: Addison-Wesley Publishing Co., 1995.

> An interesting and readable treatise on organizational strategy. The authors' ideas are strongest in the discussions of strategy in innumerable organizations.

Waldrop, M. Mitchell. *Complexity: The Emerging Science at the Edge of Order and Chaos.* New York: Simon & Schuster, 1992.

> Waldrop's book traces the history of the ascendance of complexity science by chronicling several of the people who founded the Santa Fe Institute, the center of today's scientific exploration of complexity and complex adaptive systems concepts. A great starting place for those readers who are less inclined to wade through the more scientific tomes.

_____. "The Trillion-Dollar Vision of Dee Hock." *Fast Company, 1997 Supplement,* pp. 57–62.

Weinberg, Gerald M. *Quality Software Management, Vol. 1: Systems Thinking.* New York: Dorset House Publishing, 1992.

_____. *Quality Software Management, Vol. 2: First-Order Measurement.* New York: Dorset House Publishing, 1993.

_____. *Quality Software Management, Vol. 3: Congruent Action.* New York: Dorset House Publishing, 1994.

_____. *Quality Software Management, Vol. 4: Anticipating Change.* New York: Dorset House Publishing, 1997.

> Any and all of Weinberg's books, probably thirty or so, are worthy reads. I was introduced to Jerry's writings in the early 1970's with *The Psychology of Computer Programming,* a classic reissued in 1998 by Dorset House Publishing. Jerry's *Quality Software Management* four-volume series deserves a read by anyone interested in the human side of our technological world.

Wheatley, Margaret J. *Leadership and the New Science: Learning About Organizations from an Orderly Universe.* San Francisco: Berrett-Koehler, 1992.

> Winner of Best Management Book of the Year (1992) from *Industry Week.* Wheatley links the concepts of complexity science to management at a conceptual level. A book on management philosophy with few practical specifics. Well-written and enjoyable.

Whitely, Richard. *The Customer-Driven Company.* Reading, Mass.: Addison-Wesley Publishing Co., 1991.

Womack, James, Daniel Jones, and Daniel Roos. *The Machine That Changed the World: The Story of Lean Production.* New York: Harper-Perennial, 1990.

> Some ideas change the very nature of competition. This book chronicles one such idea, telling the story of the Japanese automotive industry's transition from an era of mass production to one of lean production. The next step in the evolution, already apparent in some industries, is the transition to mass customization.

Zachary, G. Pascal. *Show-Stopper.* New York: The Free Press, 1994.

> The story of Windows NT development. An example, in my opinion, of adaptive development at work.

Index